Macmillan/McGraw-Hill Science

EXPLORING SPACE

Asteroids and comet between Jupiter and Mars

AUTHORS

Mary Atwater
The University of Georgia
Prentice Baptiste
University of Houston
Lucy Daniel
Rutherford County Schools
Jay Hackett
University of Northern Colorado
Richard Moyer
University of Michigan, Dearborn
Carol Takemoto
Los Angeles Unified School District
Nancy Wilson
Sacramento Unified School District

Macmillan/McGraw-Hill School Publishing Company
New York Chicago Columbus

CONSULTANTS

Assessment:

Janice M. Camplin
Curriculum Coordinator, Elementary Science
Mentor, Western New York
Lake Shore Central Schools
Angola, NY

Mary Hamm
Associate Professor
Department of Elementary Education
San Francisco State University
San Francisco, CA

Cognitive Development:

Dr. Elisabeth Charron
Assistant Professor of Science Education
Montana State University
Bozeman, MT

Sue Teele
Director of Education Extension
University of California, Riverside
Riverside, CA

Cooperative Learning:

Harold Pratt
Executive Director of Curriculum
Jefferson County Public Schools
Golden, CO

Earth Science:

Thomas A. Davies
Research Scientist
The University of Texas
Austin, TX

David G. Futch
Associate Professor of Biology
San Diego State University
San Diego, CA

Dr. Shadia Rifai Habbal
Harvard-Smithsonian Center for Astrophysics
Cambridge, MA

Tom Murphree, Ph.D.
Global Systems Studies
Monterey, CA

Suzanne O'Connell
Assistant Professor
Wesleyan University
Middletown, CT

Environmental Education:

Cheryl Charles, Ph.D.
Executive Director
Project Wild
Boulder, CO

Gifted:

Sandra N. Kaplan
Associate Director, National/State Leadership
Training Institute on the Gifted/Talented
Ventura County Superintendent of Schools Office
Northridge, CA

Global Education:

M. Eugene Gilliom
Professor of Social Studies and Global Education
The Ohio State University
Columbus, OH

Merry M. Merryfield
Assistant Professor of Social Studies and Global Education
The Ohio State University
Columbus, OH

Intermediate Specialist

Sharon L. Strating
Missouri State Teacher of the Year
Northwest Missouri State University
Marysville, MO

Life Science:

Carl D. Barrentine
Associate Professor of Biology
California State University
Bakersfield, CA

V.L. Holland
Professor and Chair, Biological Sciences Department
California Polytechnic State University
San Luis Obispo, CA

Donald C. Lisowy
Education Specialist
New York, NY

Dan B. Walker
Associate Dean for Science Education and Professor of Biology
San Jose State University
San Jose, CA

Literature:

Dr. Donna E. Norton
Texas A&M University
College Station, TX

Tina Thoburn, Ed.D.
President
Thoburn Educational Enterprises, Inc.
Ligonier, PA

Macmillan/McGraw-Hill School Division
10 Union Square East
New York, New York 10003

Printed in the United States of America

ISBN 0-02-274271-9 / 5

6 7 8 9 VHJ 99 98 97 96 95 94

A U.S. space shuttle blasting off

Mathematics:

Martin L. Johnson
Professor, Mathematics Education
University of Maryland at College Park
College Park, MD

Physical Science:

Max Diem, Ph.D.
Professor of Chemistry
City University of New York, Hunter College
New York, NY

Gretchen M. Gillis
Geologist
Maxus Exploration Company
Dallas, TX

Wendell H. Potter
Associate Professor of Physics
Department of Physics
University of California, Davis
Davis, CA

Claudia K. Viehland
Educational Consultant, Chemist
Sigma Chemical Company
St. Louis, MO

Reading:

Jean Wallace Gillet
Reading Teacher
Charlottesville Public Schools
Charlottesville, VA

Charles Temple, Ph.D.
Associate Professor of Education
Hobart and William Smith Colleges
Geneva, NY

Safety:

Janice Sutkus
Program Manager: Education
National Safety Council
Chicago, IL

Science Technology and Society (STS):

William C. Kyle, Jr.
Director, School Mathematics and Science Center
Purdue University
West Lafayette, IN

Social Studies:

Mary A. McFarland
Instructional Coordinator of
Social Studies, K-12, and
Director of Staff Development
Parkway School District
St. Louis, MO

Students Acquiring English:

Mrs. Bronwyn G. Frederick, M.A.
Bilingual Teacher
Pomona Unified School District
Pomona, CA

Misconceptions:

Dr. Charles W. Anderson
Michigan State University
East Lansing, MI

Dr. Edward L. Smith
Michigan State University
East Lansing, MI

Multicultural:

Bernard L. Charles
Senior Vice President
Quality Education for Minorities Network
Washington, DC

Cheryl Willis Hudson
Graphic Designer and Publishing Consultant
Part Owner and Publisher, Just Us Books, Inc.
Orange, NJ

Paul B. Janeczko
Poet
Hebron, MA

James R. Murphy
Math Teacher
La Guardia High School
New York, NY

Ramon L. Santiago
Professor of Education and Director of ESL
Lehman College, City University of New York
Bronx, NY

Clifford E. Trafzer
Professor and Chair, Ethnic Studies
University of California, Riverside
Riverside, CA

STUDENT ACTIVITY TESTERS

Jennifer Kildow
Brooke Straub
Cassie Zistl
Betsy McKeown
Seth McLaughlin
Max Berry
Wayne Henderson

FIELD TEST TEACHERS

Sharon Ervin
San Pablo Elementary School
Jacksonville, FL

Michelle Gallaway
Indianapolis Public School #44
Indianapolis, IN

Kathryn Gallman
#7 School
Rochester, NY

Karla McBride
#44 School
Rochester, NY

Diane Pease
Leopold Elementary
Madison, WI

Kathy Perez
Martin Luther King Elementary
Jacksonville, FL

Ralph Stamler
Thoreau School
Madison, WI

Joanne Stern
Hilltop Elementary School
Glen Burnie, MD

Janet Young
Indianapolis Public School #90
Indianapolis, IN

CONTRIBUTING WRITER

Fred Schroyer

EXPLORING SPACE

Lessons **Themes**

Activities!

EXPLORE

TRY THIS

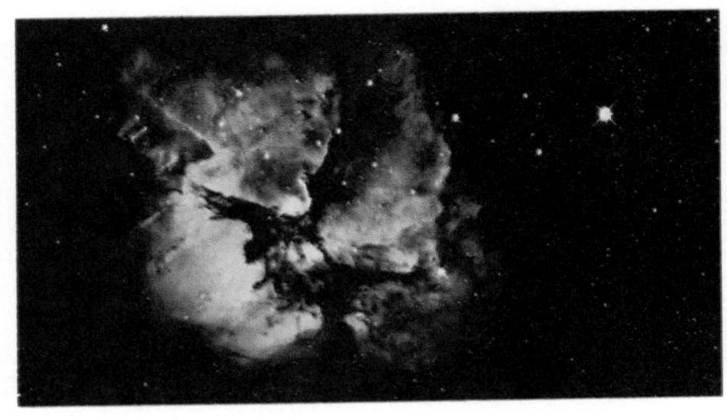

Features

Links

Departments

Theme  SYSTEMS and INTERACTIONS

EXPLORING SPACE

"Twinkle, twinkle, little star,
how I wonder what you are."

The next clear night before you go to bed, look at the sky. Do you wonder what you are seeing? People have been trying to explain what they see in the sky for a very long time. What is up there? What is space?

Minds On! In your ***Activity Log*** on page 1, draw and color what you think space looks like. Include the things that you think are in space. ●

Astronomy (ə stron´ ə mē) is the study of the stars, planets, and other bodies found in space. Astronomy is a very old science. As long as 5,000 years ago, the first astronomers in Egypt, Babylonia, and China observed space and saw patterns in the way the moon and the stars moved. But, they had a problem. Things look different far away than they really are, and people had no way of seeing things in space up close. Also, the patterns that people observed gave them the wrong idea about the patterns that really exist. Think about how things you see in space appear to you. Stars appear as tiny points of light. The moon looks like a basketball in the sky. The sun seems to circle around Earth. Earth looks flat and feels like it's holding still. You can't always trust your eyes!

If you want to learn more about space, today you have many resources to use. If you are between 10 and 14, you could even go to Space Camp in Florida or Alabama and spend a week doing the same kind of training activities as the astronauts.

People's curiosity about space has led them to invent new technology to see and measure space better. We have learned more and more about what is in space and how it functions. We've invented telescopes, rockets, and spacecraft to explore the sky and test our ideas. We have also begun to understand that Earth is part of a system, the **solar system.** The sun is the center of that system which includes Earth and eight other planets and their moons. The solar system is only one part of a larger system, a galaxy of stars. The galaxy that the Earth is in is only one of many galaxies that make up the universe.

Many things have been discovered about space, but there are still many more questions to answer. Astronomers are studying space and finding out new facts about the system in which we live. You may have heard about the discoveries of new galaxies or new objects in space. In this unit, you'll explore some of these things and how we know about them.

At Space Camp you learn to build and launch rockets. You use the same equipment that astronauts use. And you can take part in a simulated, or pretend, space mission.

Science in Literature

Although most of us cannot journey to space, we can learn about it through reading. Books offer lots of exciting ideas about space and space exploration.

The Macmillan Book of Astronomy
by Roy A. Gallant.
New York:
Macmillan, 1986.

Did you know that organisms on Earth live at the bottom of an ocean of air? That our moon has holes? That the light of stars goes out? What would happen if Earth's sun, which is a star, flickered out? This book will take you on a trip through the universe that will answer these questions and more. It might change the way you view the solar system.

A Wrinkle in Time
by Madeleine L'Engle.

New York:
Farrar, Straus and Giroux, Inc., 1962.

This award-winning book is the story of the adventures in space and time of three children—Meg, her brother, and a friend. They are in search of their father, who disappeared while doing secret government work on the tesseract problem. What's a tesseract? Well, it is a wrinkle in time. Read this book to join in the adventures of these three children as they travel through space.

Other Good Books To Read

To Space and Back
by Sally Ride, with Susan Okie

New York: Lothrop,
Lee & Shepard Books, 1986.

Can you imagine blasting off in a rocket and floating in midair while circling hundreds of miles above Earth? With photographs that make you feel like you are in a space shuttle yourself, reading this book will really make you feel that you have been to space and back!

Her Seven Brothers
by Paul Goble.

New York: Bradbury Press, 1988.

Goble's story is of a Native American girl's trip to the north country to find her seven brothers. It is a retelling of a Cheyenne legend of the creation of the Big Dipper.

The Big Dipper and You
by E.C. Krupp.

New York: Morrow Junior Books, 1989.

The constellation, the Big Dipper, is important for more than just the images people see in it. The Big Dipper can help us understand our sky.

Voyager to the Planets
by Necia H. Apfel.

New York: Clarion Books, 1991.

Voyager and *Voyager 2*, two spacecraft that the United States developed, flew closer to the outer planets than any other spacecraft. This is the dramatic story of the data and discoveries that these spacecraft sent back to Earth.

Destination: Space!

Rocket launch

When our senses don't give us enough information to understand something, people invent technology to extend them. Technology has allowed people to see, hear, taste, smell, and feel a little part of space.

Suppose you have a wrapped box. You must figure out what's inside, but you can't open it. Do you look it over? Do you pick it up to see how it feels? Do you shake it, listen, and sniff it? You are using your senses to gather data about the box.

But suppose this doesn't give you enough data. Could you help your eyes see into the box by x-raying it? Could you aid your sense of touch by weighing the box? Could you help your ears by listening to the box with a doctor's stethoscope? And could you boost your sense of smell with a machine that analyzed gases from the box?

This is how we learn about an object. We use our senses to explore it. When our sensors aren't sensitive enough, we use technology to extend them. And if we can't travel to where the thing is, we use technology to "transport" our sensors there.

Minds On! How could you throw a ball to Australia? How might you feed a bird as it flies over your school? If you wanted to do these things, what would you have to invent? In your ***Activity Log*** on page 2, design and explain a way to do these things. ●

In the next activity, you'll play rocket designer and discover some problems scientists have doing things in space.

EXPLORE

Activity!

Balloon Rocket to Mars!

Scientists extend our senses by sending instruments into space using rockets. A rocket is a vehicle that is moved by hot gases. The gases are discharged from the bottom of the rocket to make the rocket move upward. The instruments are attached to the top of the rocket and launched into space.

Launching a rocket to a planet is like trying to feed a bird as it flies over your school! The distances are very far, and the planets are moving. In this activity you'll be a rocket designer. You'll build and test a balloon rocket to carry an object to Mars.

What You Need

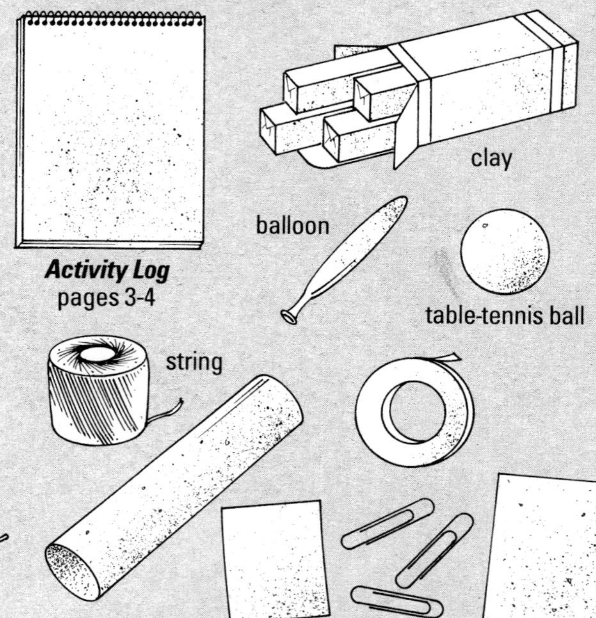

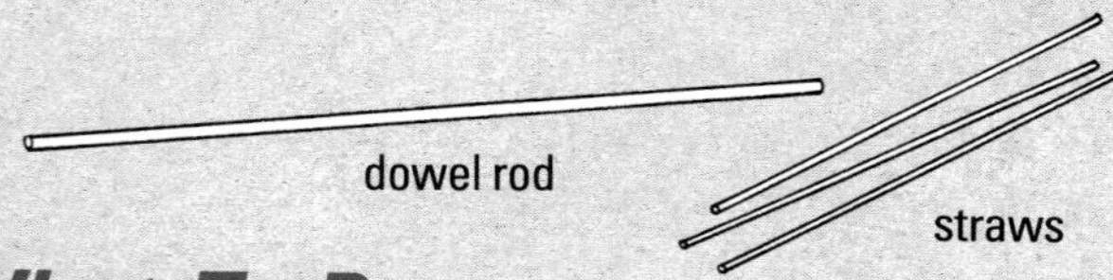

What To Do

1 Using the materials, design a balloon rocket that can carry the table-tennis ball to Mars. We'll make it easy and pretend Mars is only 3 m from your launch site.

2 Draw the plan of your rocket design in your ***Activity Log.***

3 Use the materials to build the rocket you design.

4 Mark your launch site with an X. Draw a 50-cm circle 3 m from the launch site to represent Mars. Test-fire your rocket.

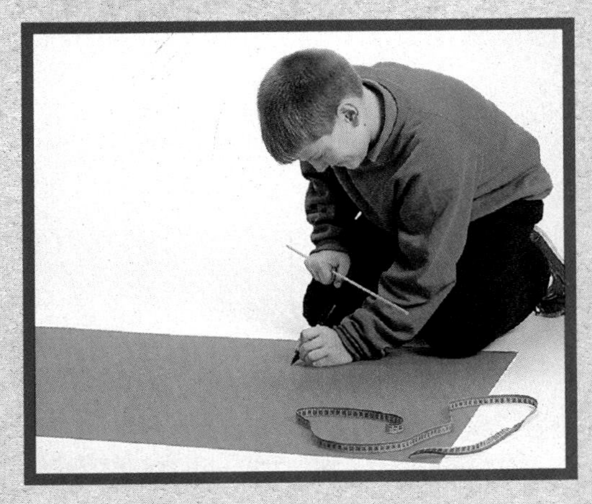

5 Modify your rocket to make it go closer to Mars. Try launching it again.

6 Test and modify your rocket three more times.

What Happened?

1. What caused your rocket to move?
2. What things affected the flight of your rocket?
3. What made it fly better or fly worse?

What Now?

1. What did you change to make your rocket more accurate? How well did it work? If you were to change your rocket more, how could you make it even more accurate?
2. Compared to people who design actual rockets, what advantages did you have in designing yours? What disadvantages?
3. Why do people build rockets?

Sensing Space

Rockets

Your activity showed how tricky it is to make a rocket go to another planet. You pretended Mars was only three meters away, instead of millions of kilometers. Your Mars was holding still, instead of moving through space. Real rockets are more difficult to design and control.

The Chinese began making gunpowder rockets to use as weapons twelve hundred years ago.

By World War II, rockets could fly many kilometers. Although rockets have often been developed for war, without them people would never have been able to break through Earth's atmosphere to get to space.

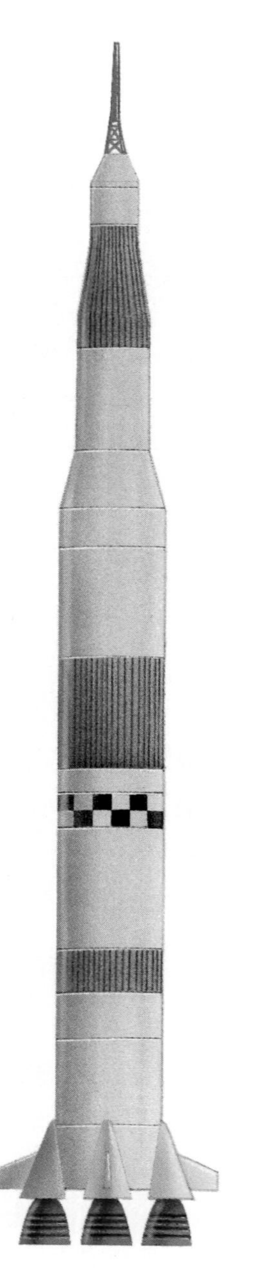

The United States space program is run by NASA (National Aeronautics and Space Administration). NASA sent the first Americans to space, put the first people on the moon, and developed the space shuttle program.

Telescopes

We must use technology to extend our senses in space because our vision is very limited. Optical telescopes are telescopes that astronomers use to make things appear closer. They gather light and allow the observer to focus the light to see objects more clearly. With a telescope, an astronomer looking into space can see the color and movement of stars. The astronomer can also see features, such as mountains, on planets.

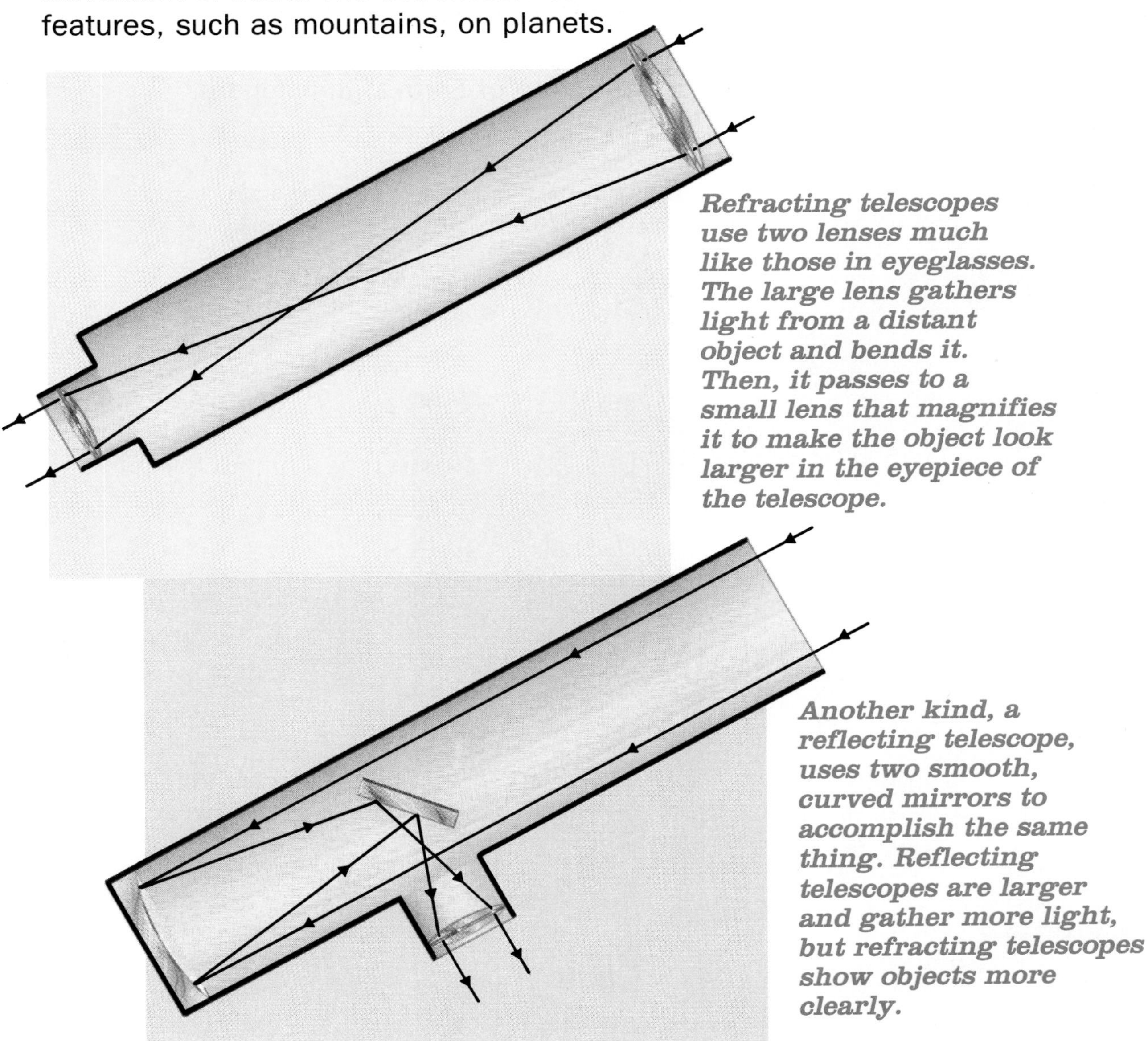

Refracting telescopes use two lenses much like those in eyeglasses. The large lens gathers light from a distant object and bends it. Then, it passes to a small lens that magnifies it to make the object look larger in the eyepiece of the telescope.

Another kind, a reflecting telescope, uses two smooth, curved mirrors to accomplish the same thing. Reflecting telescopes are larger and gather more light, but refracting telescopes show objects more clearly.

Although light from distant stars comes to Earth day and night, telescopes must be used at night because sunlight blocks starlight. Clouds also make it difficult to see with optical telescopes. Therefore, many large telescopes are located in clear, desert climates, such as Arizona, or they're located above the clouds on mountain tops.

One of the best locations for a telescope is in space itself. The Hubble Space Telescope was launched by rocket in 1990. It is orbiting Earth, but it is focused on space. It can see seven times deeper into space than a telescope on Earth, because it is free of Earth's atmosphere.

Radio telescopes gather radio waves rather than light waves from space. They can be used during the day and when the sky is cloudy. Scientists need many radio telescopes to accomplish the same thing as an optical telescope.

The Hubble Telescope in the U.S. Space & Rocket Center museum is a full-scale exhibit that explains how the orbiting, reflecting telescope works.

Satellites

The Space Age began in 1957 when the Soviet Union used a rocket to launch* Sputnik 1*, the first artificial* satellite *(sat´ ə līt´), into space.

A satellite, such as the Hubble Telescope, is an object that revolves around a larger body in space. There are two types of satellites—natural ones, like the moon, and artificial ones, such as *Sputnik 1.*

To put a satellite into space, the rocket carrying it has to go fast enough to overcome the pull of Earth's gravity. The satellite enters a path or orbit around Earth which scientists are able to control. Satellites are held in place by **gravity** which is the force that pulls objects to one another, no matter how small or far apart they are. Earth's gravity is the force that keeps you, the moon, and many artificial satellites from floating away.

Much of the information we get from satellites tells us about Earth or helps us communicate better on Earth. Environmental scientists study pollution, geologists look for oil, and agricultural scientists observe crops with satellites. Communication satellites transmit radio and television programs around Earth. The important exception is the Hubble Space Telescope which is looking into space rather than at Earth.

Weather satellites watch the atmosphere from above so we can forecast the weather, including the location of dangerous storms.

Mapmakers draw precise maps from satellite pictures.

Space Probes

Minds On! Pretend you are exploring a dark cave with no source of light. Feeling your way, you come to a cliff. How far down does it drop? And what's at the bottom? You drop a pebble and listen. Three seconds later, splash! What information did your probe send back?

Pioneer 10

Scientists do the same thing with space probes. A **space probe** is a spacecraft launched by a rocket that travels close to distant planets or other objects in space. Like the pebble you pretended you dropped in the cave, probes send their information and don't come back. Try the next activity to see how difficult sending probes can be.

We extend our senses with space probes. Radio and TV transmitters on space probes send back pictures and data. Space probes don't orbit or return to Earth.

TRY THIS **Activity!**

How Does Distance Affect Accuracy?

How can you hit a moving target with a ball? What factors do you need to keep in mind?

What You Need

jar, meter tape, table-tennis ball, chalk, paper, pencil, *Activity Log* pages 5–6

Choose a starting point. Make marks on the floor with chalk at 25 cm, 50 cm, and 100 cm from the starting point. Have a partner hold the jar at 25 cm. From the 0 mark, toss the ball 10 times. Record how many times it goes in the jar. Repeat for 50 cm and 100 cm. Next, repeat by throwing the ball while your partner moves the jar rapidly from side to side.

Repeat the experiment with your partner tossing the ball. Combine your data with the entire class and draw a bar graph in your ***Activity Log***. At what distance was your throwing accuracy best? At what distance was it the worst? How did the moving target affect your accuracy? Why is it difficult to send probes to an exact location?

The farther away an object is and the faster it moves, the harder it is to hit. Imagine trying to gently land a spacecraft on a planet! It's now possible to do this because we've invented powerful rockets, computers, and radio controls. Space probes are tracked with huge antennas located in the United States, Australia, and Spain. Computers are used to communicate with probes.

Just as building one balloon rocket showed you ways to make a better one, so early flights taught NASA scientists better ways to build probes. Probes are sending back detailed pictures and valuable information.

NASA's early Mariner probes flew by Mercury, Venus, and Mars.

TRY THIS **Activity!**

Do-It-Yourself Mars Probe

Design a probe to send back data about Mars. Decide what data from Mars you would like to collect.

What You Need

various construction materials,
***Activity Log* pages 7–8**

Make a sketch of your probe in your ***Activity Log*** and label its parts. Then construct a model of your probe. You could use construction paper, boxes, tubes, or other materials.

Space Shuttles

Although technology allows us to gather information without traveling to space, there are many things that we can learn only by going there. The U.S.S.R. and the United States have sent many people to space in several space flight programs. In the United States, this included the *Mercury*, *Gemini*, and *Apollo* programs. The latest development is the reusable spacecraft, the ***space shuttle***. Launched by giant rockets, the shuttle orbits Earth while astronauts do their work.

When the astronauts' work is completed, they are able to pilot the shuttle back to Earth to be used again.

Shuttle astronauts launch satellites and repair them. They also use the shuttle as an observatory for solar and Earth measurements.

Focus on Environment

We're Even Littering Space!

Artificial satellites don't last forever. Pieces of old rockets, dead satellites, and bits of metal are whirling around Earth. A tiny paint fleck chipped a space shuttle windshield and cost 50 thousand dollars to fix! The space shuttle *Discovery* had to steer away from a truck-sized piece of a U.S.S.R. rocket that came too close. Space station *Freedom* will have special armor to protect it. How can we clean up this junkyard in the sky?

Social Studies Link

Steps Into Space

Guion Bluford, Jr., first African American astronaut

1232 Chinese develop gunpowder rockets

1609 Galileo (gal ə lā´ ō), an Italian scientist, uses telescope to study planets and stars

1840 American astronomer takes first photos of the moon

1903 Wright Brothers, United States inventors, fly first engine-powered airplane

1937 United States astronomers use first radio telescopes

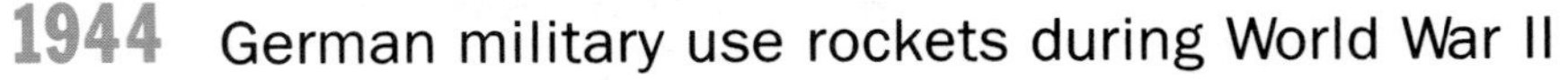

1944 German military use rockets during World War II

1957 Soviet Union orbits *Sputnik 1*, first artificial satellite

1961 Yuri Gagarin, a Soviet cosmonaut, is first person in space

1969 *Apollo 11* astronauts from the United States land on moon

1977 *Voyager 2*, United States probe, flies by Jupiter, Saturn, Uranus, Neptune

1981 First United States space shuttle, *Columbia*, takes off from Florida and lands in California

1990 United States and the European Space Agency launch Hubble Space Telescope into orbit

Sally Ride, first U.S. female astronaut

On a blank sheet of paper, lay out a time line using the dates above. Using your ruler, make the line 25 centimeters long. Allow 2.5 centimeters for each 100 years. Write in the events in the correct place on the line. Attach your time line to your ***Activity Log*** page 9. Why is it so difficult to fit in all the events at the end of the line? Do you know why there is such a gap between 1232 and 1609?

Should People or Machines Explore the Planets?

What's the best way to see something? Go there! That is why many scientists want to visit Mars and other planets. The problem is that it's very dangerous and expensive to send people into space. Some people believe that the billions of dollars spent by NASA would be better spent on other programs to help people with their problems. The explosion of the space shuttle *Challenger* in 1986 was proof for some people that space travel is too dangerous and that it's better to send probes and robots to space. They say we can get just as much information with machines.

Do you think we should explore the planets with people or machines? Make a list of pros and cons (advantages and disadvantages) of each kind of exploration in your ***Activity Log*** page 10. If you had to choose one type, which would it be? Share your list with your class. Did your classmates agree with your choice?

Apollo *spacecraft*

Apollo *moon landing*

Learning From Space Travel

You have learned about technology that has helped people understand space and space travel. Technology has been developed to overcome problems with space travel, too.

Health Link

Medicine and Weightlessness

Astronauts in space experience the same sensation of **weightlessness** you feel when an elevator starts or when you ride on a roller coaster. Being weightless is fun, but it can cause an astronaut's muscles to weaken, bones to lose calcium, and face to swell. Also, they may experience motion sickness. Doctors who treat astronauts must solve these problems before astronauts can make long trips to planets and live in space stations. Experiments and research being done on many shuttle flights may help people on Earth who have similar health problems.

At Space Camp, the microgravity simulator is designed to give trainees a realistic feeling of the reduced gravity Apollo astronauts experienced on the moon.

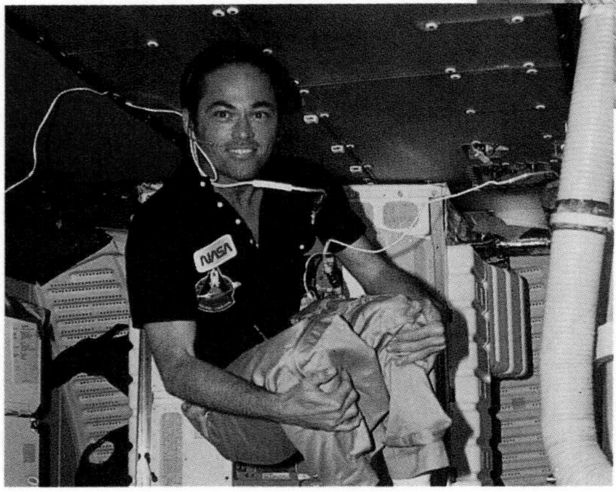

Weightless astronaut

Five Degrees of Freedom or 5DF allows the Space Camp trainee to simulate the environment of space. Movement is allowed in five directions (except up and down). The 5DF gives the trainee hands-on experience in working in a weightless environment.

Try this activity to see what happens to water in a container when it falls.

TRY THIS **Activity!**

Weightless Water

Will water run out of a cup that is falling?

What You Need

pencil, paper cup, water, sliding board, *Activity Log* page 11

Over a sink, poke a hole in the side of a large paper cup or milk carton with a pencil. Fill the cup with water. What happens to the water? Now fill the cup again and put your finger over the hole. Take the cup to the playground and climb up the sliding board steps. Drop the cup. Does any water come out while the cup is falling? Why? Draw a diagram of what happened in your ***Activity Log.*** Water inside a falling cup is similar to astronauts riding in a spacecraft. Why?

Focus on Technology

Are Your Sneakers From Space?

Because of space travel, we have invented many new things to cope with space problems such as being weightless, and living and working in a different environment. Many of the athletic shoes that you can buy today have a layer of cushioning that was developed for walking on the moon. You may be wearing a pair right now. You also might be wearing eyeglasses with plastic lenses that won't shatter. These were also designed for wear in space. You enjoy other space inventions, too.

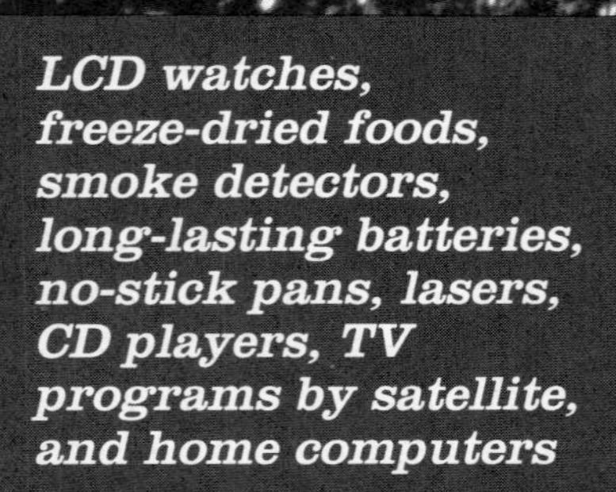

LCD watches, freeze-dried foods, smoke detectors, long-lasting batteries, no-stick pans, lasers, CD players, TV programs by satellite, and home computers

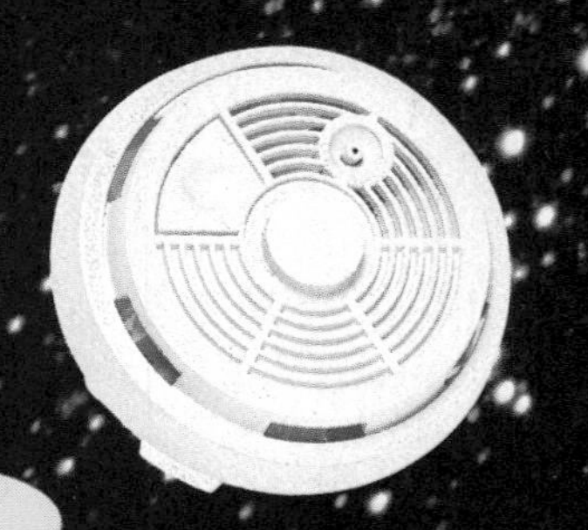

More than 2,500 new inventions have been developed by the space program. If we weren't exploring space, we wouldn't have these things! Make a list in your ***Activity Log*** on page 12 of all the things you have in your home or school that rely on inventions from space exploration.

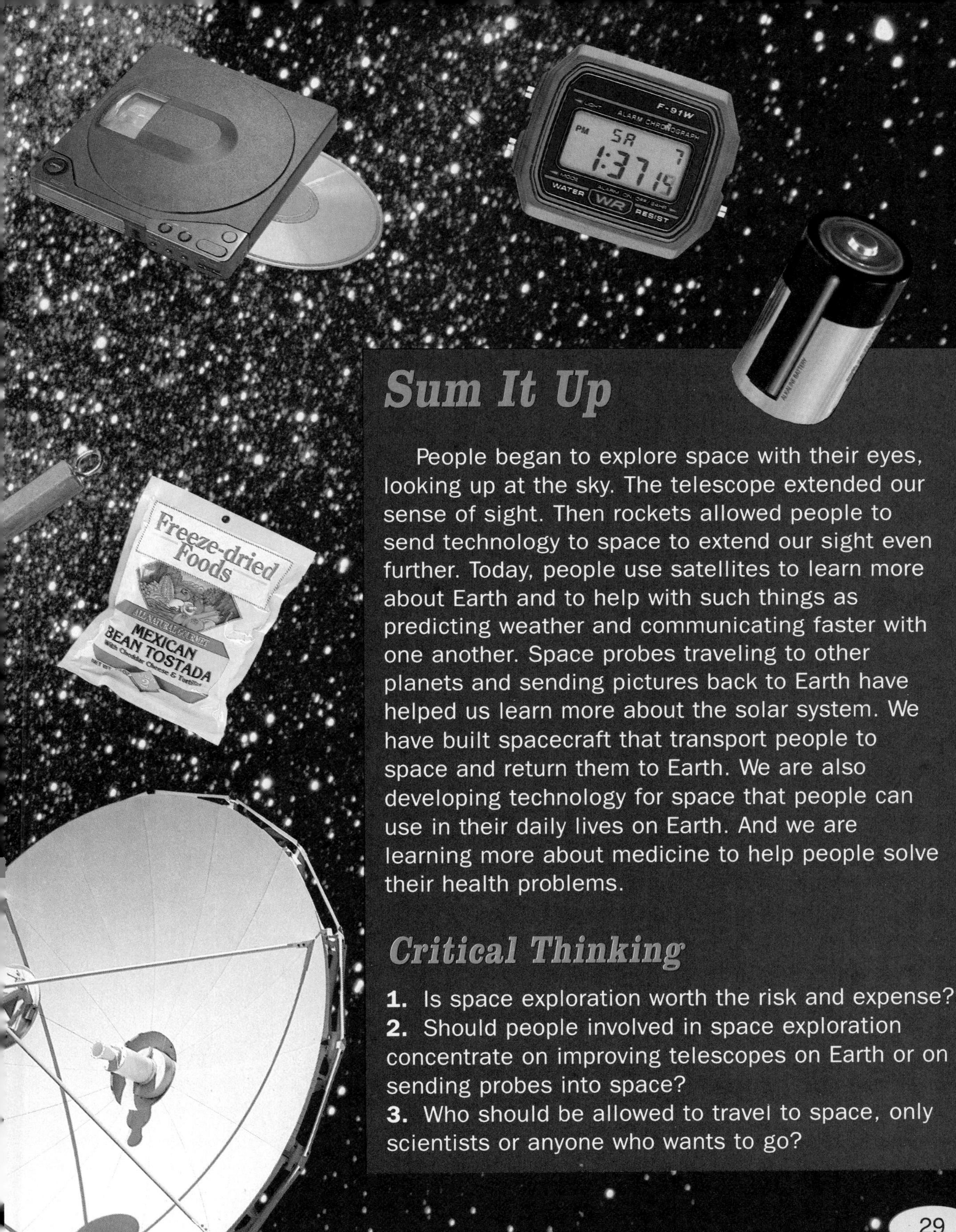

Sum It Up

People began to explore space with their eyes, looking up at the sky. The telescope extended our sense of sight. Then rockets allowed people to send technology to space to extend our sight even further. Today, people use satellites to learn more about Earth and to help with such things as predicting weather and communicating faster with one another. Space probes traveling to other planets and sending pictures back to Earth have helped us learn more about the solar system. We have built spacecraft that transport people to space and return them to Earth. We are also developing technology for space that people can use in their daily lives on Earth. And we are learning more about medicine to help people solve their health problems.

Critical Thinking

1. Is space exploration worth the risk and expense?
2. Should people involved in space exploration concentrate on improving telescopes on Earth or on sending probes into space?
3. Who should be allowed to travel to space, only scientists or anyone who wants to go?

Patterns in the Sky

Sunrise over the Nile River

Our lives today follow patterns in the sky. Day after day, season after season, year after year, we pattern our lives on the movement of Earth and the moon around the sun. We also use patterns in the stars and constellations to find our way.

Imagine living five thousand years ago along the Nile River in Egypt. One hot July morning you awaken to unusual sounds. People are excited about spotting the brightest star in the sky, Sirius (Sir′ ē əs). It's rising just ahead of the sun. Everyone knows this event in the sky happens just before the Nile River floods, bringing fertile soil. The rising of Sirius, the brightest star we can see, was so important in ancient Egypt that it marked the New Year.

Minds On! How do patterns in the sky still affect life today? Write the ways you can think of in your ***Activity Log*** on page 13. ●

For thousands of years, people in all cultures have studied the patterns of the sun, moon, and stars. Hopi and Zuni Native Americans were able to predict seasons by observing and marking the sun's position. Others, including the Mayans of Mexico, made calendars to predict events. The patterns told them when to plant crops or go hunting. Patterns in the stars helped travelers navigate. Polynesian Islanders were able to sail from Tahiti to Hawaii by using the stars. People even put shadows to work! In the next activity, you'll explore how they used shadows.

EXPLORE

Activity!

Sun Shadows

Can you use the sky to tell time?

What You Need

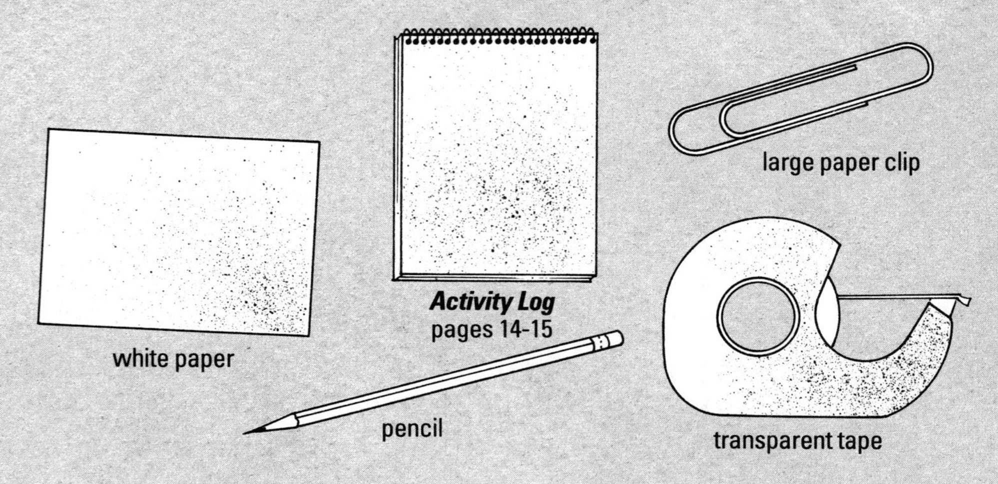

What To Do

1 Bend one side of the paper clip to make a stand.

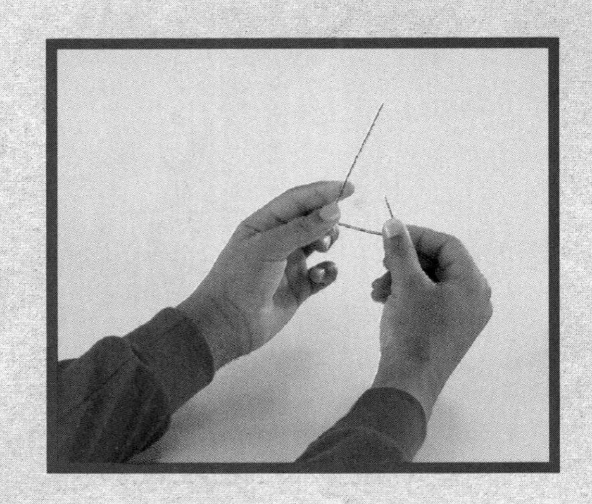

2 Stick the paper clip through the center of the paper.

3 Tape the base to the back side of the paper.

4 Place the paper on a flat, smooth surface where the sun will shine on it. Make sure it won't be disturbed all day. Use tape or a pencil to mark the paper's exact location on the ground or table.

5 At 10:00 A.M., carefully trace the paper clip's shadow. Put a heavy dot on the tip of the tracing. Repeat at 11:00 A.M., 12:00 P.M., and 1:00 P.M. Label your tracings with the times.

6 Predict where you think the tip of the shadow would be at 9:00 A.M. and at 2:00 P.M. Mark these predicted points on the paper.

7 Return the paper to its marked location the next day. Trace the actual shadows at 9:00 A.M. and 2:00 P.M.

What Happened?

1. On the 1st day, what did you observe? What did the shadow do?
2. When was the shadow longest? When was it shortest?
3. How much did the shadows change from the 1st day to the 2nd day?

What Now?

1. How could you use a shadow to tell the time of day?
2. Does the change in the shadow length from day to day create a problem with telling time? Explain your answer.
3. How accurate were your predictions? What could you do to make them more accurate?
4. How could you tell time on a cloudy day?
5. What patterns in the sky affect shadows?

Sky Patterns

Earth and the Sun

The activity used a paper clip's shadow to show how Earth steadily rotates. The Earth rotates or spins around, much like a toy top, once every 24 hours. The sun appears to be in different positions as Earth moves. This is how we define a day. The sun comes into view at sunrise. Shadows are very long then. Shadows grow shorter until midday, when Earth has rotated so the sun is highest in the sky. Then, as Earth turns, shadows lengthen until sunset.

*The pattern of shadows cast by the sun tells time. The activity was making a "shadow clock," or **sundial**. Sundials work because the sun shines steadily, making shadows, and Earth rotates steadily, moving the shadows. They always point in the same direction. People in all cultures have used this daily pattern of Earth's rotation to tell time.*

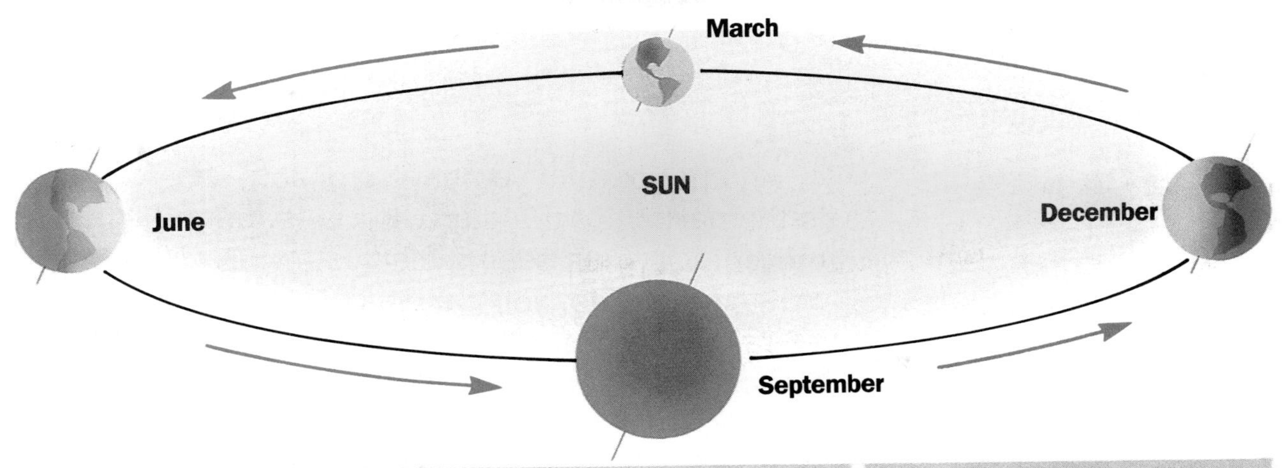

Noon-June 21

Noon-March 21/September 21

Noon-December 21

It takes a year, or 365 1/4 days, for Earth to travel in its orbit around the sun. Because Earth is tilted, the rays of the sun do not hit Earth at the same angle at all times of the year.

Shadows are shorter in summer in the Northern Hemisphere when the sun's light hits there more directly and are longer in winter when the sun's light is less direct.

The seasons are reversed in the Southern Hemisphere.

One of the most difficult ideas for people to understand was that Earth is moving and the sun is standing still. In 1543, Polish astronomer Nicolaus Copernicus (kə pûr´ ni kəs) wrote a book suggesting that the sun is at the center of the solar system with the planets revolving around it. His idea did not fit what people saw. As telescopes improved observations, his hypothesis proved to be correct.

Stars and Constellations

Another pattern you have probably observed is the pattern of the stars. In cities, bright lights prevent you from seeing many stars. But in the dark countryside, you can see one thousand to three thousand stars. With a telescope or binoculars, you can see thousands more. Only a few bright stars, such as Sirius, stand out.

People have always imagined that they could draw lines to connect the stars in the sky to make pictures. These artificial pictures, or visual star patterns, are **constellations** (kon´ stə lā´ shənz).

Minds On! With your eyes closed, quickly make 20 dots on page 16 in your ***Activity Log.*** These are stars. Do you see a picture in your stars? This is your own constellation. Name it! Now, show your stars to two classmates. Do they see different pictures? If people from Japan or Zambia looked at your stars, would they see different pictures? ●

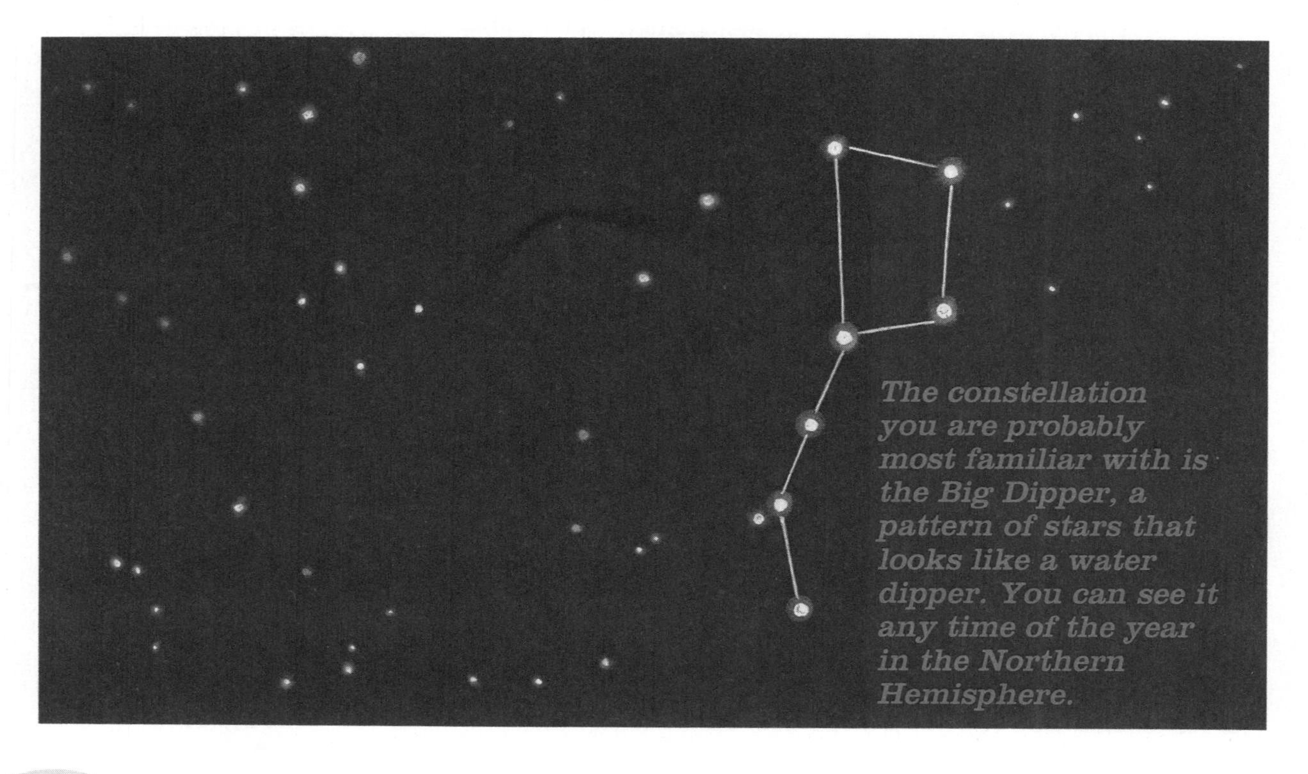

The constellation you are probably most familiar with is the Big Dipper, a pattern of stars that looks like a water dipper. You can see it any time of the year in the Northern Hemisphere.

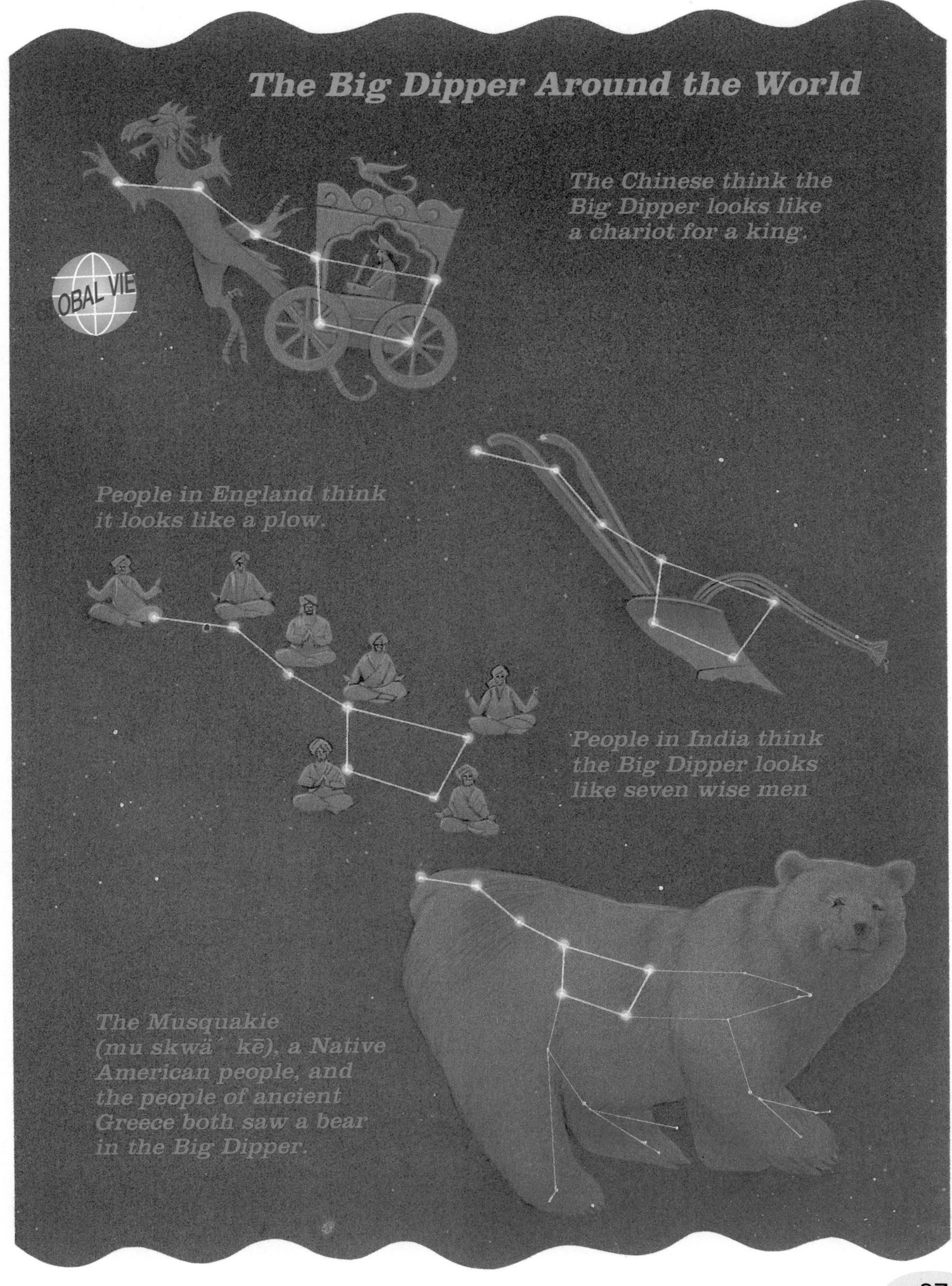
The Big Dipper Around the World
The Chinese think the Big Dipper looks like a chariot for a king.
People in England think it looks like a plow.
People in India think the Big Dipper looks like seven wise men
The Musquakie (mu skwä´ kē), a Native American people, and the people of ancient Greece both saw a bear in the Big Dipper.

An important set of constellations are those around the North Pole known as the circumpolar constellations. They can be seen any time of the year.

They include Cassiopeia (kas´ ē ə pē´ ə), the queen,

Cepheus (sə fē´ əs), the king,

and Draco (drā´ cō), the dragon.

People in ancient Greece thought they saw pictures of a queen, a king, and a dragon in these constellations.

Find these constellations on your star map and see if you can see a queen, a king, or a dragon.

Another important constellation in the winter sky is Orion, the hunter. It is easy to recognize because it has two very bright stars at opposite corners of the constellation — Betelgeuse (be´ təl jüz´) and Rigel.

Could you use your star map to find these constellations outdoors? The following activity will help you recognize them.

TRY THIS Activity!

Shoe Box Constellation

You will make a model of a constellation that will help you recognize constellations in the night sky.

What You Need

shoe box with lid, piece of black construction paper, push pin, scissors, transparent tape, cardboard, Activity Worksheet (star map), pencil, *Activity Log* page 17

Cut a 1-cm hole in one end of the box and a 7 x 13-cm hole in the opposite end. Find a constellation on the star map. Place the construction paper and cardboard under the star map and carefully use the push pin to poke a hole through each star in your constellation. Remove the construction paper and tape it over the large opening and put the lid on the box. Hold it up to a bright light and peek through the hole. What do you see? Trade boxes with your classmates and find their constellations on the star map. Draw your constellation in your ***Activity Log.***

Safety Tip: Be careful not to push the pin into your finger.

The stars you see at night are very far away. Stars in constellations may be farther from each other than some are from us.

If you look at constellations at the same time every night, they seem to move a little from night to night. But are they moving, or is Earth? Remember, Earth revolves steadily around the sun. These daily movements make the constellations appear to move. As Earth orbits the sun, some constellations gradually move out of sight and new ones come into view, following a yearly pattern or cycle.

12:00 midnight

Even more noticeable than the constellations' yearly cycle is their nightly change in position. Earth revolves around the sun, but Earth also rotates.

Because Earth rotates, the stars in the constellations "move" each night. Depending on your location, they appear to rise and set, just like our sun appears to rise and set.

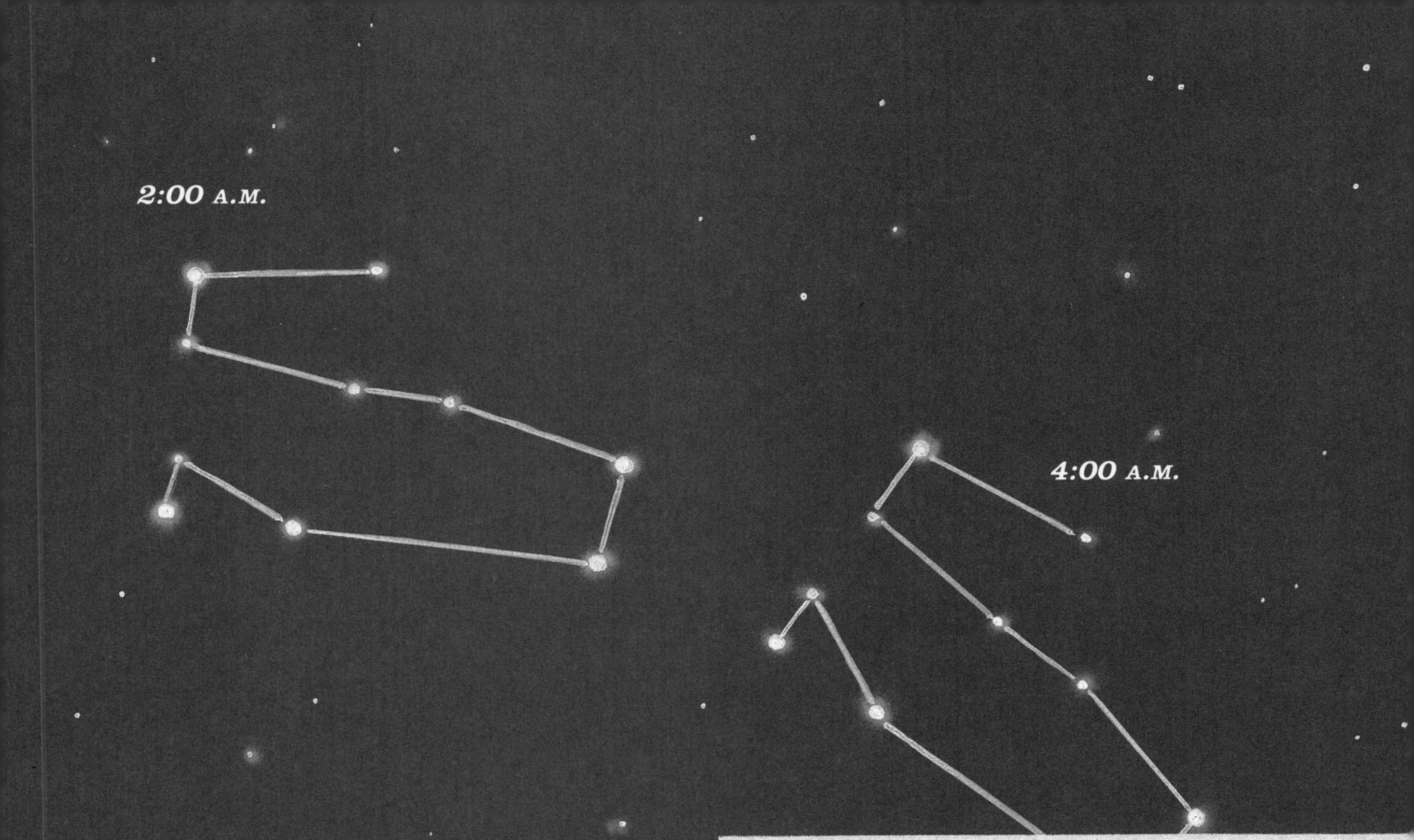

2:00 A.M.

4:00 A.M.

TRY THIS **Activity!**

Star Movement

At home tonight, try this activity that shows how stars appear to move in the sky.

What You Need

starry night, pencil, *Activity Log* page 18

Stand under the edge of a building or under the large branch of a tree. Find a bright star in the sky and note its location with respect to the tree or building. Watch the star for at least 15 mins. What happened to it? In your ***Activity Log,*** explain why this happens.

Finding Your Way

People use constellations to navigate at night. They are like highway signs in the sky. Because constellations are easy to pick out among all the stars, people also use them to find individual stars.

How can you find one star among the thousands you see? The same way ancient people did—start with a familiar constellation. Do the Try This Activity to locate directions.

TRY THIS Activity!

There's N! Where Are E, W, and S?

By knowing where the North Star is, you can locate other directions.

What You Need

Activity Worksheet (star map), piece of white paper, pencil, *Activity Log* page 19

To find the North Star, use the star map and find the Big Dipper. The stars that form the front of the Big Dipper's bowl point to the North Star. Once you have located the North Star, you know where north is. Draw the Big Dipper and the North Star in your ***Activity Log.***

Sailors navigating by the stars

Along with clocks, navigators use an instrument called a **sextant.** A sextant measures the angle of a star or the sun from the horizon.

The North Star is very important in navigation in the Northern Hemisphere because it is almost directly above Earth's North Pole. Its location in the sky never changes. The South Pole does not have a polar star, but it has a constellation, the Southern Cross, that points in the direction of the South Pole.

The big problem for navigators is finding their way east and west. As Earth rotates, the constellations rise and set. If you are a navigator and you are moving east or west, the constellation you are watching may disappear. You know the direction you are going, but not how far in that direction you have gone. The star you watched in the Try This Activity on page 41 was moving about one inch every 15 minutes, relative to the tree or roof. When accurate clocks were invented, the problem of navigating east and west was solved. If people knew what time it was, they had a way to measure how far east or west they were from any point.

Planetarium Director

A planetarium is a place where projectors are used to project the image of stars and other objects in space on a dome-shaped screen. The projector can make objects move. Constellations rise and set, and spacecraft move through the sky. Planetarium shows are an exciting way for people to learn about space.

Many planetariums are in museums. A person who manages a planetarium is responsible for developing and conducting shows for visitors, managing equipment, and supervising people who work in the planetarium. The director may also be responsible for museum exhibits about space. To become a planetarium director, you must take courses in astronomy, and you need to learn about museum management as well.

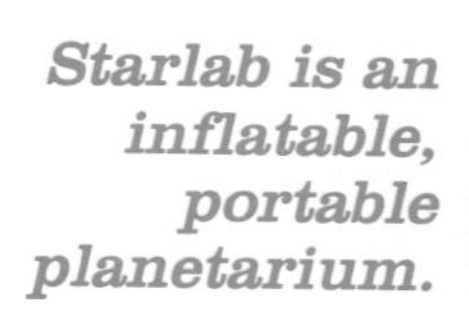

Starlab is an inflatable, portable planetarium.

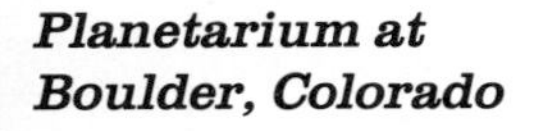

Planetarium at Boulder, Colorado

Inside a planetarium

Sum It Up

Patterns in the sky have had much influence on people's actions. People have learned about the sun and the stars by observing patterns repeating over time. Before they understood that Earth revolved around the sun and rotated, they knew that the sun and star patterns were reliable. They could use these patterns to make predictions about the seasons and to find their way from one place to another. People were able to develop calendars, tell time, and plant crops using the sun. All cultures have imagined constellations in the stars and have used them to navigate and to make star maps.

Critical Thinking

1. How can you use the sun and stars to tell the time of day or the time of year?
2. Why do you think people made up stories about the constellations?
3. Why do you think people used to think that Earth was not moving?

Stars in the Galaxies

Constellation Orion

Constellations are spread across the night sky. You can imagine you see animals and people in the constellation patterns, but what about the individual stars that make up the constellations? In this lesson you will learn about the life cycle of stars, about galaxies, and about how we measure the enormous distances in space.

While they look like very small points of light in the night sky, some stars, like Betelgeuse in the constellation Orion, are huge. Astronomers usually focus their telescopes on individual stars rather than on constellations.

Minds On! What do you think astronomers see when they look at stars? In your ***Activity Log*** on page 20, draw a star. ●

Did you put points on your star? Do stars really have points? They look much smaller than our sun and moon, but are they? Are all the stars you see the same size and color? What else is out there in space? The night sky is filled with questions!

Tonight, notice all the lights you see. Watch how car lights change when they get closer and farther away. If you can see stars, notice how they are alike and how they are different. Take a flashlight outside and see how far the beam can shine. In the next activity, you'll use flashlights to learn more about the stars.

EXPLORE

Activity!

How Do Distance and Size Affect Brightness?

Using flashlights, you will learn how the distance and size of stars affects their brightness.

What You Need

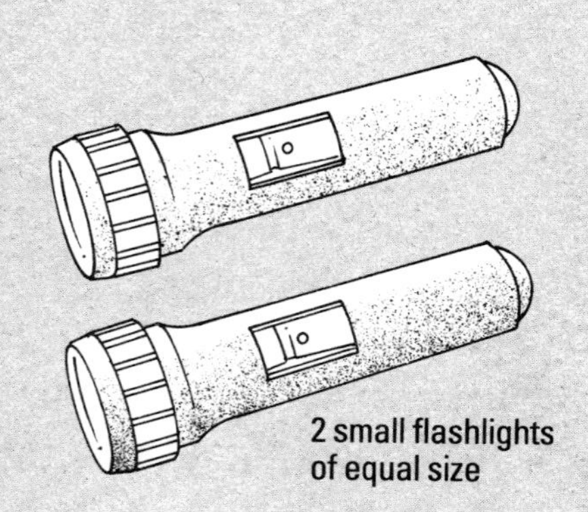

2 small flashlights of equal size

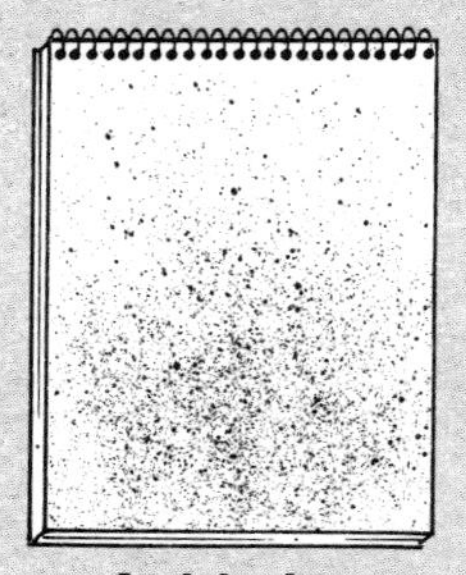

Activity Log pages 21-22

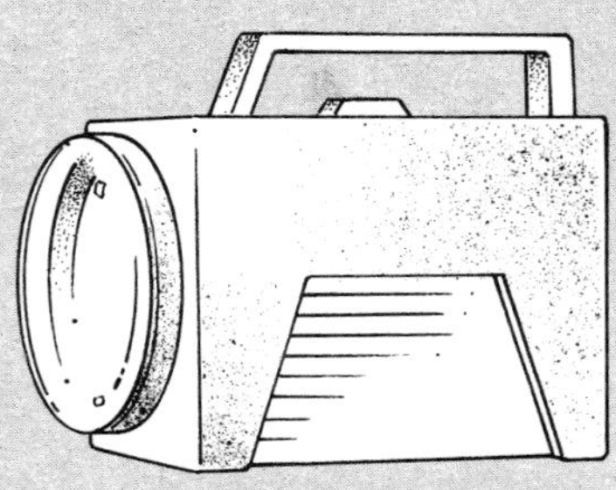

one large flashlight

What To Do

1 Compare the brightness of the large flashlight with one of the smaller flashlights by turning them on and looking at them. Record your observations in your ***Activity Log.***

2 Compare the brightness of the two smaller flashlights. Record your observations in your ***Activity Log.***

3 Find a way to make the large flashlight look the same brightness as one of the smaller flashlights, using distance. Describe in your ***Activity Log*** what you found.

4 Now, find a way to make one of the smaller flashlights look dimmer than the other, using distance. Describe in your ***Activity Log*** what you found.

What Happened?

1. Where did the observer have to be to see the large and small flashlight as having the same brightness?
2. Where did the observer have to be to see one of the small flashlights as dimmer than the other?
3. What is the connection you found between brightness and distance?

What Now?

1. How could you apply your observations of the flashlights to stars and their brightness?
2. What could make a smaller, dimmer star appear brighter than a larger, brighter star?
3. Write a rule about how bright a star looks, how bright it is, and how far away it is. How could you test your rule?

Star Light, Star Bright

What Is a Star?

In the Explore Activity, you made a model of stars. Flashlights the same size gave off the same light energy. But the closer one looked brighter. Flashlights of different sizes gave off different light energy. But the smaller one looked as bright as the larger one when it was closer. Starlight functions in the same way.

Stars are huge balls of gas. Their gravity causes hydrogen atoms to join together and make helium. This makes stars very hot. They give off energy in many forms, including heat, light, and radio waves.

Our sun is a star. Like all other stars we know of, it is spherical in shape. The sun is only about 150 million kilometers (93 million miles) away. That is 546 thousand times closer than the brightest star, Sirius.

Our sun is only a medium-sized star, but it is 109 times bigger in diameter than Earth. If it were hollow, you could stuff a million Earths inside it!

Astronomers use the sun as a guide to understanding other stars.

Astronomers cooperate with scientists in other fields, particularly physics, chemistry, and mathematics. Together they are able to analyze starlight and calculate the temperature, mass, size, and composition of stars. They can compute distances to stars based on the size of Earth and measurements of changes in the position of stars in the sky.

We know that our sun is yellow, but stars come in a variety of colors. There are blue, white, orange, red, and yellow stars. A star's color shows how hot its surface is and how large it is. The hottest stars are blue, and the coolest are red.

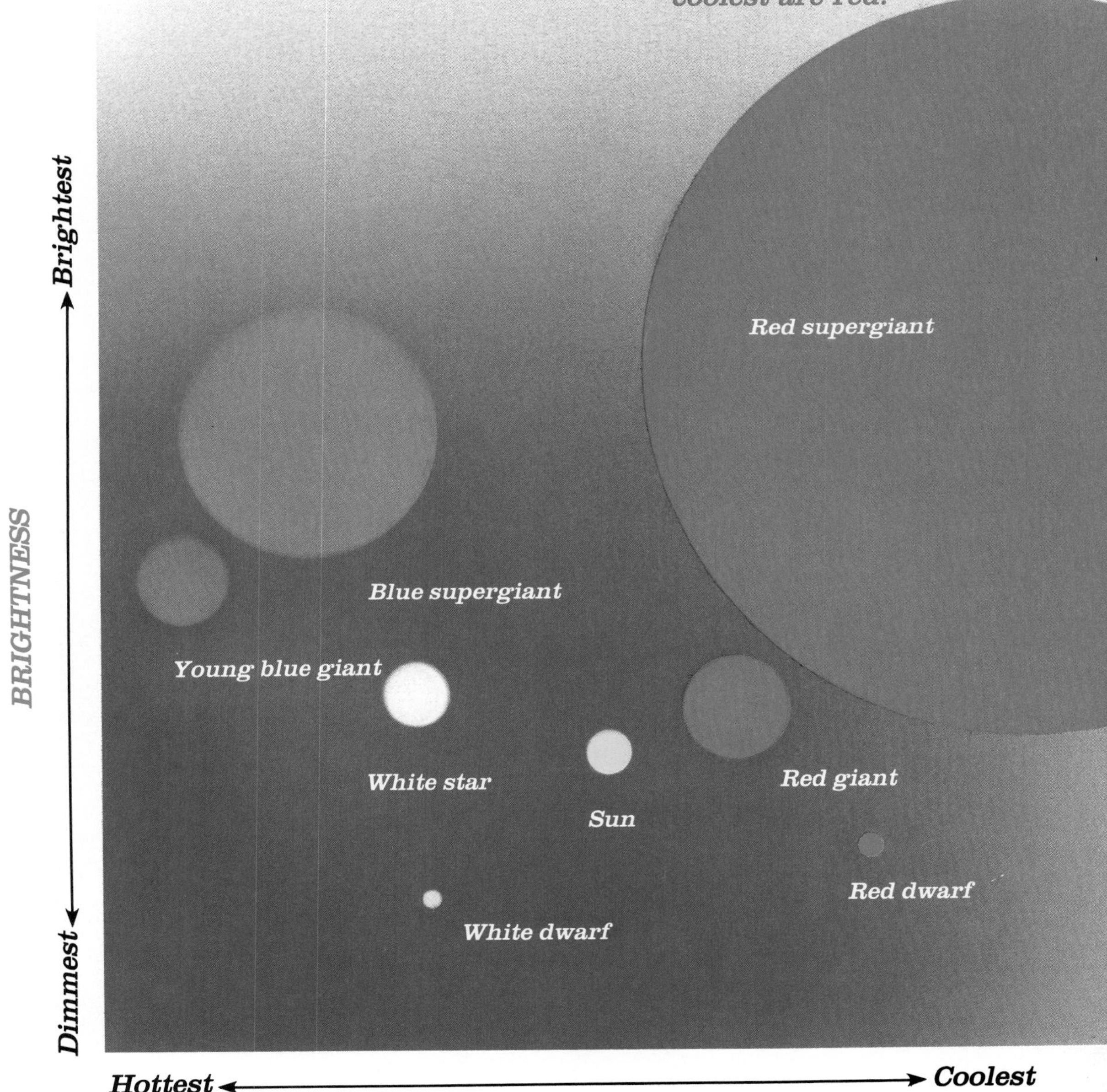

Star Cycle

You may think that stars are permanent, but they're not. Over billions of years, they slowly form, grow, and die. The life of many stars follows a pattern.

1 *A great cloud of hydrogen gas and dust called a* ***nebula*** *(neb´ yə lə) may be the start of a star. Gravity slowly pulls the gas and dust closer together, forming the core. The more compact the matter gets, the hotter the star gets.*

2 *When the core temperature reaches about 1 million degrees Celsius, some of the hydrogen combines to form helium and a large amount of energy is produced. The energy is released into space as light, heat, and other forms of energy.*

3 *Eventually most of the core hydrogen is changed to helium. Then, helium begins changing into carbon, and the star expands. As it expands, its outer layer cools. The star becomes a red giant.*

4 *What happens next depends on the mass of the star. Stars the size of the sun and smaller throw off the outer layers to become a nebula, and the core cools. The core becomes a white dwarf, giving off the heat it has stored.*

*If the star has a mass three times larger than the sun, it becomes a supergiant. These stars may explode, producing a **supernova** (sü´ pər no´ və). Supernovas are spectacular events. The most recent was seen in 1987. The material remaining after the supernova can form nebulae, neutron stars, pulsars, or black holes depending on their size.*

Supernova

Crab Nebula

Nebulae are one of the remains of a supernova. The Crab Nebula in the constellation Taurus is the remains of a supernova recorded in A.D. 1054 by Chinese astronomers.

You know what pattern a star's life follows. In the following activity, try to draw what you think the star cycle of our sun will be.

TRY THIS

Activity!

Our Sun From Birth to Death

What You Need

crayons or markers, *Activity Log* pages 23–24

In your ***Activity Log,*** draw the star cycle of our sun. Color and label each stage. Mark where our sun is in the sequence.

Galaxies

The thousands of stars that you see on a clear night may seem like a lot of stars to you, but there are many more stars than just those you can see. In space there are many galaxies. A **galaxy** is a huge group of stars, gas, and dust. There are billions of stars spread out across each galaxy. Our solar system is in the Milky Way Galaxy. On a clear night away from city lights, you can see part of it as a faint, milky path across the sky. The many galaxies in space are very distant from each other.

Astronomers are not in a good position in the Milky Way to observe its structure, but we know that the sun is orbiting the galaxy just as Earth orbits the sun. The Milky Way is part of a cluster of galaxies that astronomers call the Local Group. The Local Group has 30 or more galaxies.

The Milky Way galaxy is spiral-shaped and rotates like fan blades. Our sun and planets are in one arm of the spiral.

Measuring the Universe

The **universe** is space and all the matter and energy in it. How vast is the universe? No one knows. If you drove a car day and night at 85 kilometers (52.8 miles) per hour, you'd reach the sun in 200 years. But if you traveled to the next nearest star, it would take over 55 billion years!

TRY THIS Activity!

How Many Paper Clips Wide Is Your Desk?

Distances in space are very large. You can see how large by measuring with paper clips.

What You Need

desk top, paper clips, pencil, *Activity Log* page 25

Measure your desktop with paper clips. Record your measurement in your ***Activity Log.*** Now, measure from your desk to the door in your classroom and record your measurement. Did you have enough paper clips? Imagine measuring your trip home from school with paper clips!

For your trip home, kilometers are much easier to use than paper clips. But in space, distances are so great that kilometers are like paper clips. They're too small. This is why astronomers use light-years to measure space distances. The distance light travels in one year is a **light-year.**

In one year, light travels 9.5 trillion kilometers (6 trillion miles). If there were a star 9.5 trillion kilometers from Earth, it would take one year for the light to reach us. By the time the star's light reached us, it would be a year old. Looking at stars is like looking back in time.

The nearest star is our sun, only 1/63,000 of a light-year away. But the second-closest star is Proxima Centauri (prok´ sə mə sen tor´ē), about 4.3 light-years away. Other stars are hundreds, thousands, and millions of light-years away.

Distance in Kilometers to Proxima Centauri

speed of light in kilometers/second =	**300,000**	km/sec.
	×60	sec. min.
speed of light in kilometers/minute =	**18,000,000**	km/min.
	×60	min./hr.
speed of light in kilometers/hour =	**1,080,000,000**	km/hr.
	×24	hrs./day
speed of light in kilometers/day =	**25,920,000,000**	km/day
	×365¼	days/yr.
speed of light in kilometers/year =	**9,467,280,000,000**	km/yr.
	×4.3	light-years
	40,709,304,000,000	km to Proxima Centauri

4.3 light years

Proxima Centauri

Literature Link

A Wrinkle in Time

Meg and her brother Charles haven't seen their father for several years. He was working for the government on a special project which may have taken him light-years away from Earth. A stranger agrees to help them overcome this great distance to find Mr. Murry. The stranger will use a "wrinkle in time." As you read *A Wrinkle in Time*, answer the following questions in your ***Activity Log*** on page 26.

- What is a tesseract? Do you think this could really occur?
- Using what you know about stars and distances in space, how far away do you think Meg's father was from home?
- Do Meg and her companions experience weightlessness while traveling in space? How is their experience different from that of astronauts?
- Do you think life-forms in other parts of the universe would experience the same emotions as Meg and her companions? Why or why not?

Math Link

News Flash from Andromeda Galaxy!

Andromeda (an drom´ ə də) is 2.25 million light-years away. If there were a big explosion in this galaxy, how soon would we know about it? Explain your answer in the ***Activity Log*** on page 27.

Sum It Up

Stars are far away and difficult to study. Astronomers use the sun as a model of other stars. Using scientific knowledge from many fields, we have found that stars have a life cycle. When a star dies, what happens to it depends upon how massive it is. It may become a supernova, pulsar, or black hole. Our sun, Earth, and many stars are tiny parts in the Milky Way galaxy, one of countless galaxies scattered throughout the universe. Galaxies are large groups of stars and gas. Distances between stars and galaxies are so great that astronomers measure them in light-years.

Critical Thinking

1. Is it likely that our sun will become a supernova? Why or why not?
2. Why do we want to know about stars that are so far away?
3. What kind of technology could we invent that would help us learn more about stars, galaxies, and the universe?

Andromeda galaxy

My Very Educated Mother Just Served Us Nine Pizzas!

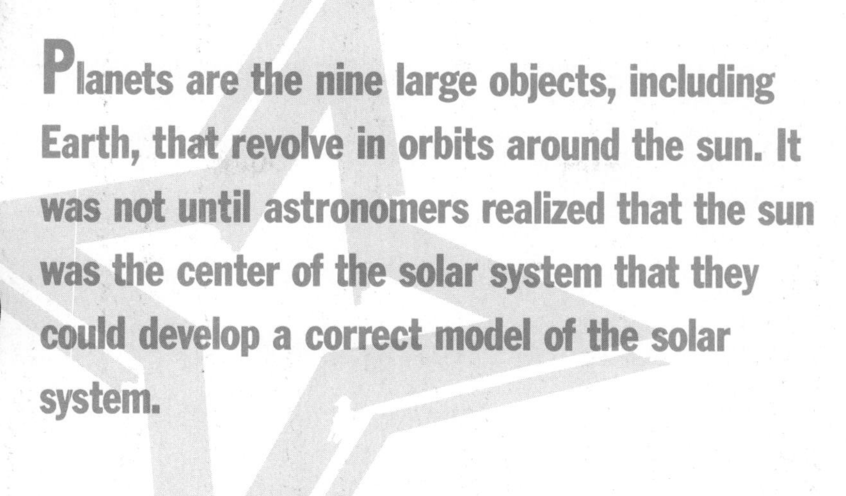

Planets are the nine large objects, including Earth, that revolve in orbits around the sun. It was not until astronomers realized that the sun was the center of the solar system that they could develop a correct model of the solar system.

As ancient people studied the stars, they noticed five that they called wanderers. These five were unusually bright and clear. Instead of moving regularly like the stars move, they seemed to have no pattern. They could be seen in different constellations at different times of the year. These "wanderers" are five of the **planets** in the solar system.

Minds On! Look at this lesson's title. To find out what it means, look at the first letter of each word—M, V, E, M, J, S, U, N, P. The silly sentence helps you remember the planets' names, in order of their distance from the sun—Mercury, Venus, Earth, Mars, Jupiter, Saturn, Uranus, Neptune, and Pluto. Anything that helps you remember something is a mnemonic (nə mon′ ik). Can you invent your own mnemonic to help remember the planets in order? Record your ideas in your ***Activity Log*** on page 28. ●

Of the nine planets, people first saw only the six closest to the sun. One of these six was Earth. They couldn't see the outer three because these planets are so far away. In the next activity, you'll see how far each planet is from the sun. Then you'll explore the solar system to see how the planets are alike and different.

Saturn

EXPLORE

Activity!

The Solar System

By making a model of the positions of the planets in the solar system, you can compare and contrast their distances from each other and from the sun.

What You Need

Activity Log
pages 29-30

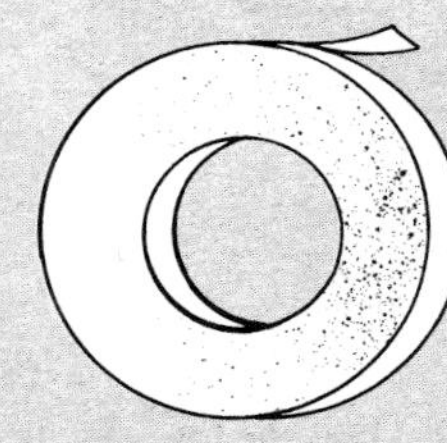

masking tape

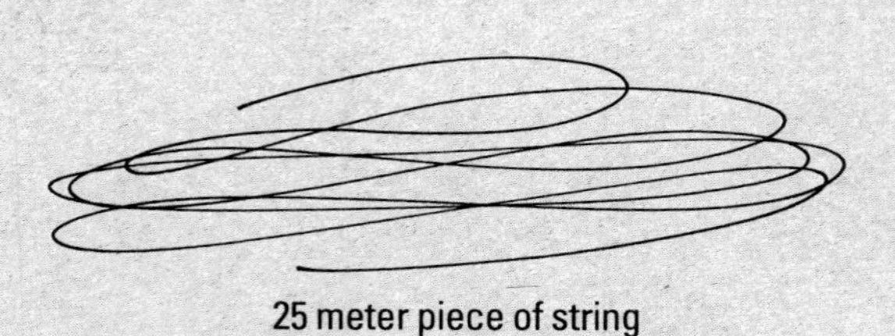

25 meter piece of string

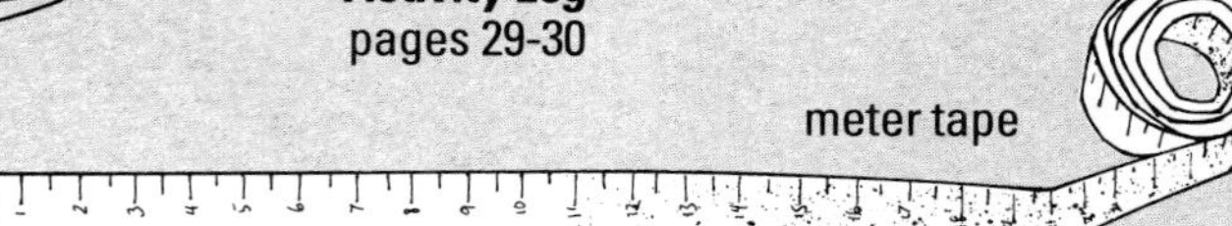

meter tape

What To Do

1 Mark the string at one end with a piece of masking tape. Write SUN on the tape.

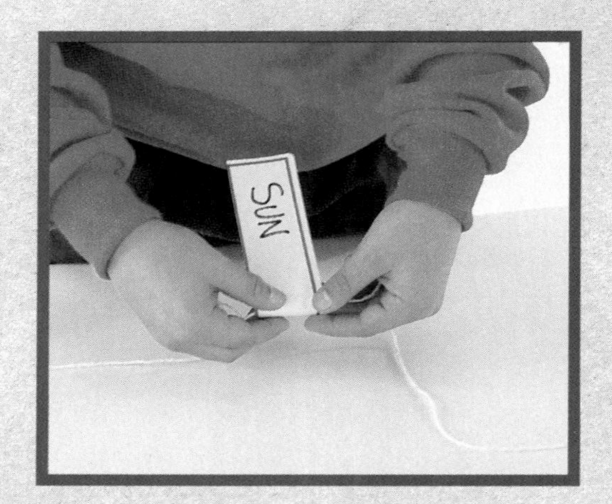

2 Measure from the SUN along the string 25 cm. Mark this point with tape, and write MERCURY on it.

3 Measure from the SUN along the string 50 cm. Mark this point with tape, and write VENUS on it.

4 Using the following information, keep measuring along the string and using the tape to label the planets. (Make each measurement from the sun, not from the last planet.)

Earth—55 cm
Mars—1 m
Jupiter—3.5 m
Saturn—6 m
Uranus—12.5 m
Neptune—19.5 m
Pluto—25 m

5 Lay the string out straight with you at the SUN end and a partner at the PLUTO end. Have another member of your group stand at EARTH.

What Happened?

1. Which two planets are the closest to each other?
2. Which planet is about halfway between the sun and Pluto?

What Now?

1. How could you reduce the model's size to fit into your classroom?
2. Use a calculator to figure out how many times farther it is from the sun to Pluto than it is from the sun to Earth. (Hint: convert all measurements to cm. Then divide the sun-to-Pluto distance by the sun-to-Earth distance.)
3. Write the names of the planets in your ***Activity Log*** in the order they are from the sun.

Our Neighbors in the Solar System

The activity showed that the nine planets are in two groups. The inner four planets are close to the sun and each other. The outer five planets are much farther from the sun and each other. Planets are spheres like stars. Like Earth and the moon, planets do not give off their own light. They shine because they reflect sunlight.

Just as our sun is a model for learning about stars, Earth is a model for learning about the planets and their moons. We are able to understand the information gathered by probes and moon landings because we can compare it to information gathered on Earth.

The Inner Planets

The four inner planets are made of rock. Most of the inner planets are smaller and hotter than the outer planets. They have changed since they were formed. In the description of each planet, years are defined in terms of Earth years, and days in terms of Earth hours and days. Remember, a year is the time it takes a planet to revolve around the sun, and a day is the time it takes a planet to rotate.

Mercury

Distance From the Sun:
57.9 million km
Year:* *88 days
Day:* *59 days
Diameter:* *4,878 km
Number of Moons:* *0
Special Features:
Mercury has the greatest variation between night and day temperatures.

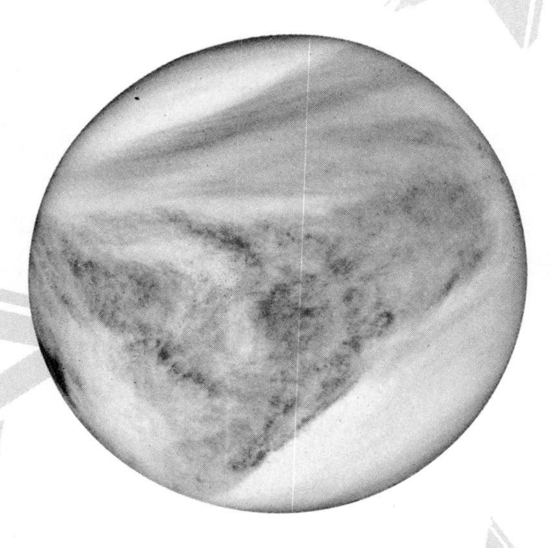

Venus

Distance From the Sun:
108.2 million km
Year: 224 days
Day: 243 days
Diameter: 12,104 km
Number of Moons: 0
Special Features:
Venus is the hottest planet because the atmosphere traps the sun's heat.

Earth

Distance From the Sun:
149.6 million km
Year: 365 days
Day: 24 hours
Diameter: 12,756 km
Number of Moons: 1
Special Features:
Earth is the only planet with liquid water.

Mars

Distance From the Sun:
227.9 million km
Year: 687 days
Day: 24½ hours
Diameter: 6,787 km
Number of Moons: 2
Special Features:
Space probes have found that Mars had liquid water at one time, but it only has ice on the surface now.

The Outer Planets

Most of the outer planets are much larger and colder than the inner planets. Most of them are "gas giants" and are made of gases such as hydrogen, helium, methane, and ammonia.

Jupiter

Distance From the Sun:
778.3 million km
Year:* *4,333 days
Day:* *9 hours
55 minutes
Diameter:* *142,800 km
Number of Moons:
At least 16
Special Features:
Jupiter has a violent atmosphere, including a large red spot that is a persistent, swirling storm.

Saturn

Distance From the Sun:
1.43 billion km
Year:* *10,759 days
Day:* *10 hours
39 minutes
Diameter:* *120,660 km
Number of Moons:
At least 21
Special Features:
Saturn is the lightest planet. It is less dense than water and could float.

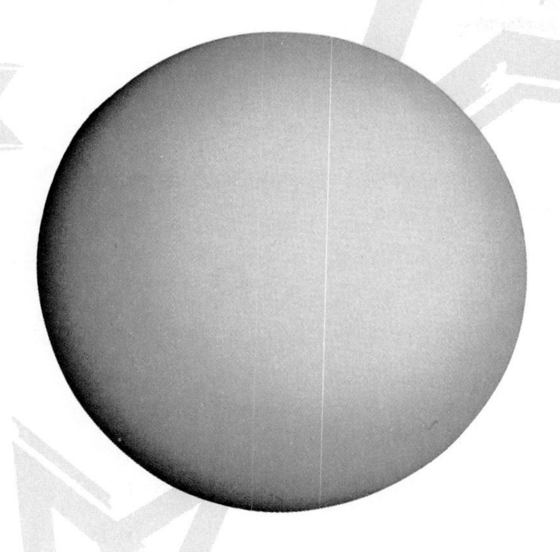

Uranus

Distance From the Sun:
2.87 billion km
Year: ***30,685 days***
Day: ***17 hours***
8 minutes
Diameter: ***51,000 km***
Number of Moons:
At least 15
Special Features:
Uranus is tilted on its axis more than any other planet.

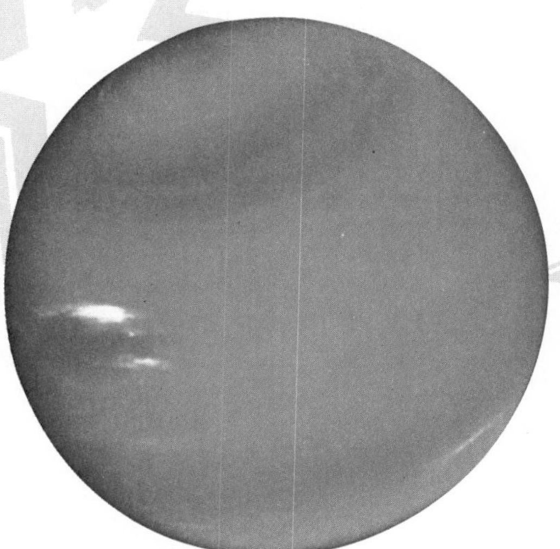

Neptune

Distance From the Sun:
4.5 billion km
Year: ***60,188 days***
Day: ***16 hours***
7 minutes
Diameter: ***49,500 km***
Number of Moons:
At least 8
Special Features:
The Great Dark Spot on Nepture is a large rotating storm, like the red spot on Jupiter.

Pluto

Distance From the Sun:
5.9 billion km
Year: ***90,700 days***
Day: ***6.39 days***
8 minutes
Diameter:
2,253 km
Number of Moons: ***1***
Special Features:
Astronomers predicted Pluto's existence before they actually found it with a telescope in 1930.

TRY THIS **Activity!**

How Do the Planets Compare?

A good way to compare information about the planets is to make a chart.

What You Need

pencil, *Activity Log* pages 31-32

Complete the chart in your ***Activity Log.*** Write in the information about each planet, and then use the chart you make to compare the planets and answer the following questions.

1. Which planet is the largest in size?
2. Which planet has the most moons?
3. Which planet has the longest year? Shortest year?
4. Which planet has the longest day? Shortest day?
5. Is there any relationship between the distance of a planet from the sun and the length of the day? Year?
6. How does a chart help you compare information about the planets?

Meteoroids, Asteroids, and Comets

Meteoroids *(mē´ tē ə roidz´) are small bits of metal and rock that orbit the sun.* ***Meteors*** *and* ***meteorites*** *are fragments of comets or asteroids that enter Earth's atmosphere. Meteors burn up in the atmosphere, while meteorites actually land on Earth's surface. Meteors are sometimes called "falling stars" because they burn in a bright streak of light as they enter the atmosphere.*

Meteorites are rare. The few we have found have been chemically tested and are a valuable source of information about the solar system. Recently, scientists have found microscopic diamonds in meteorites. They think the diamonds are fragments of stars older than the age of the solar system, 4.6 billion years.

Willamette Meteorite found in Oregon in 1902

***Asteroids** (as´ tə roidz′) are little, rocky "micro-planets" that orbit the sun, mostly between Mars and Jupiter. The largest asteroid, Ceres (sîr′ ēz), is about one-third the size of Earth's moon.*

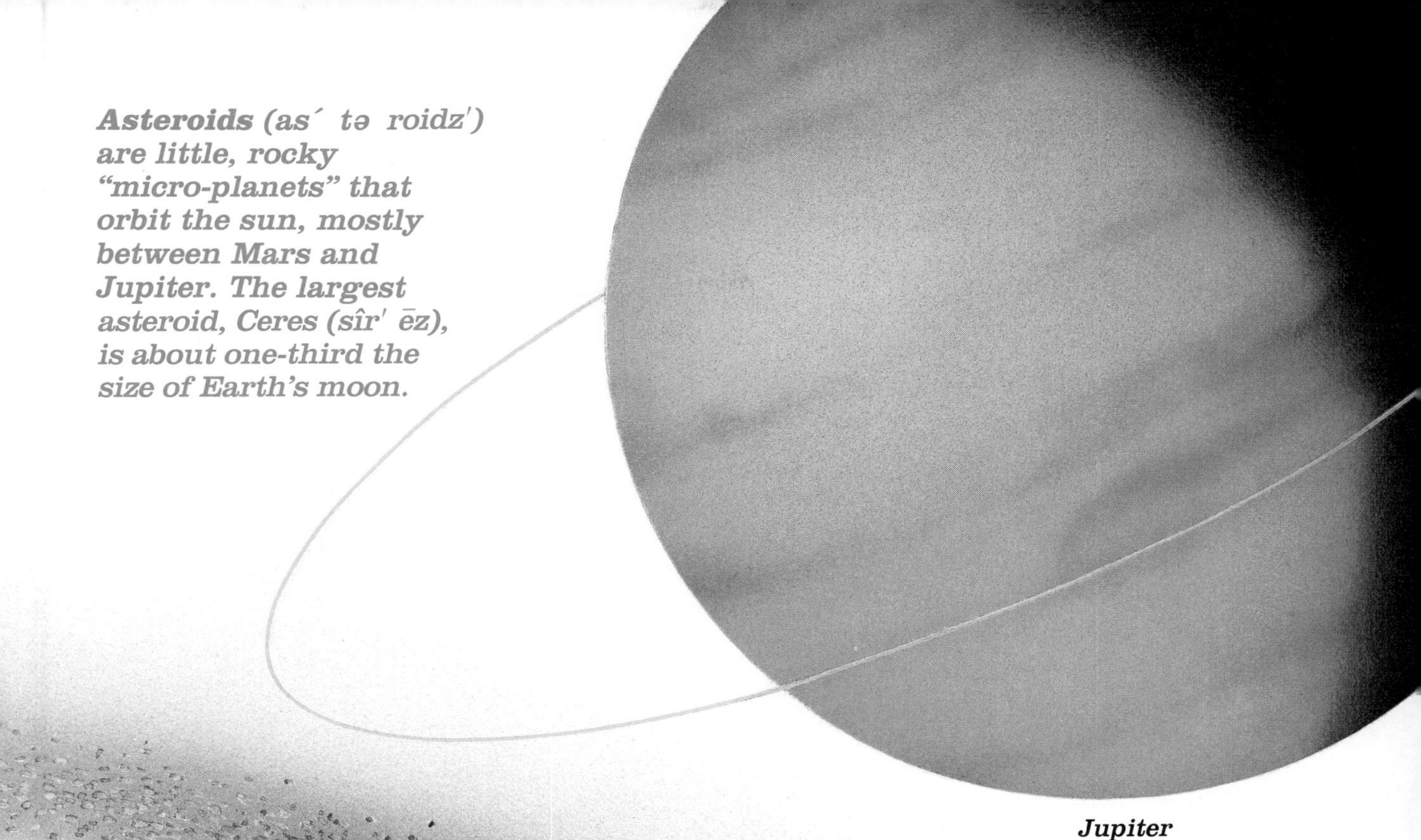

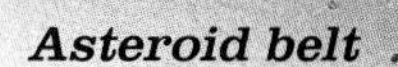

***Comets** are made of frozen gases, rock pieces, and dust. They orbit the sun in long, narrow orbits. Halley's (hal´ ēz) comet is one of the short-period comets and most easily seen on Earth. It is visible every 76 years. The last time was in 1985–1986. When can we expect to see it again?*

Mars

Comet

Art not to scale

How Did the Solar System Begin?

Scientists are not sure how the solar system began, but they have developed models to explain it. Many of these models are nebular models. Nebular models say that the sun and the planets began forming from a cloud of dust and gas about 5 billion years ago.

Hydrogen gas, helium gas, and dust were slowly pulled together by gravity. In the nebula, gravity pulled the gas and dust toward the center. The more material that went to the center, the denser it became. The cloud began to rotate, and it flattened into a wheel-shape.

About 5 billion years ago, the cloud's center became extremely dense. The hydrogen atoms started combining, making helium and releasing energy. The cloud's center became a star, our sun.

The rotation from the cloud continued, and the new planets orbited the sun. Why didn't the planets just fly off into space? Again, it's gravity. The gravitational pull between the sun and the planets held them in orbit.

Not all of the gas and dust reached the center. Some continued to rotate in the outer parts of the nebula. Gravity pulled the dust and gas particles together. The larger ones attracted more dust, growing bigger and bigger. This slowly formed the planets.

Scientists have to test how well the nebular models fit what they know about the planets, moon, sun, and other objects in space. Do the models explain what we can see and what we learn from probes? The nebular theories are accurate in some ways, but not in others. For instance, the models do not explain why the sun doesn't spin faster than it does. From what we know about Earth, the sun should spin faster and the outer planets should spin more slowly. There is still a lot of work for future astronomers before they are sure how the solar system began.

What's Out There?

You know there are billions of stars. Do any of these stars have planets that revolve around them as in our solar system? Astronomers are searching for another star with a planet revolving around it. One way to find a planet is to observe pulsars, spinning neutron stars. If a pulsar sends radio signals that are not normal, it may indicate that a planet is disrupting the signals.

Two radio telescopes in the United States search for radio signals from space 24 hours a day, every day. Also, there are people listening for radio signals from space using an invention that can tune in 8 million radio signals to see if they might be from someone trying to communicate with us. This project is called SETI which stands for Search for Extraterrestrial (ek strə tə res´ trē əl) intelligence.

Array of radio telescopes

Minds On! Suppose there is life on other planets. Do you think the life on other planets would resemble life as we know it on Earth? Would there be similar plants and animals? If you had a chance to communicate with life on another planet, what would you say? In your ***Activity Log*** on page 33, write out ten questions you would like to ask a person from another planet. ●

Would life on other planets look like E.T.?

Language Arts Link

Mars Diary

In the future, NASA plans to send the first people to Mars. The astronauts for that mission may be chosen from people who are in elementary school today. It could be you! Using what you have learned about the planets, write a diary of a day of exploration on Mars. In your diary, explain why you are exploring Mars rather than some other planet, how you got to Mars, and what the planet is like. For additional Mars information, you can use *The Macmillan Book of Astronomy.*

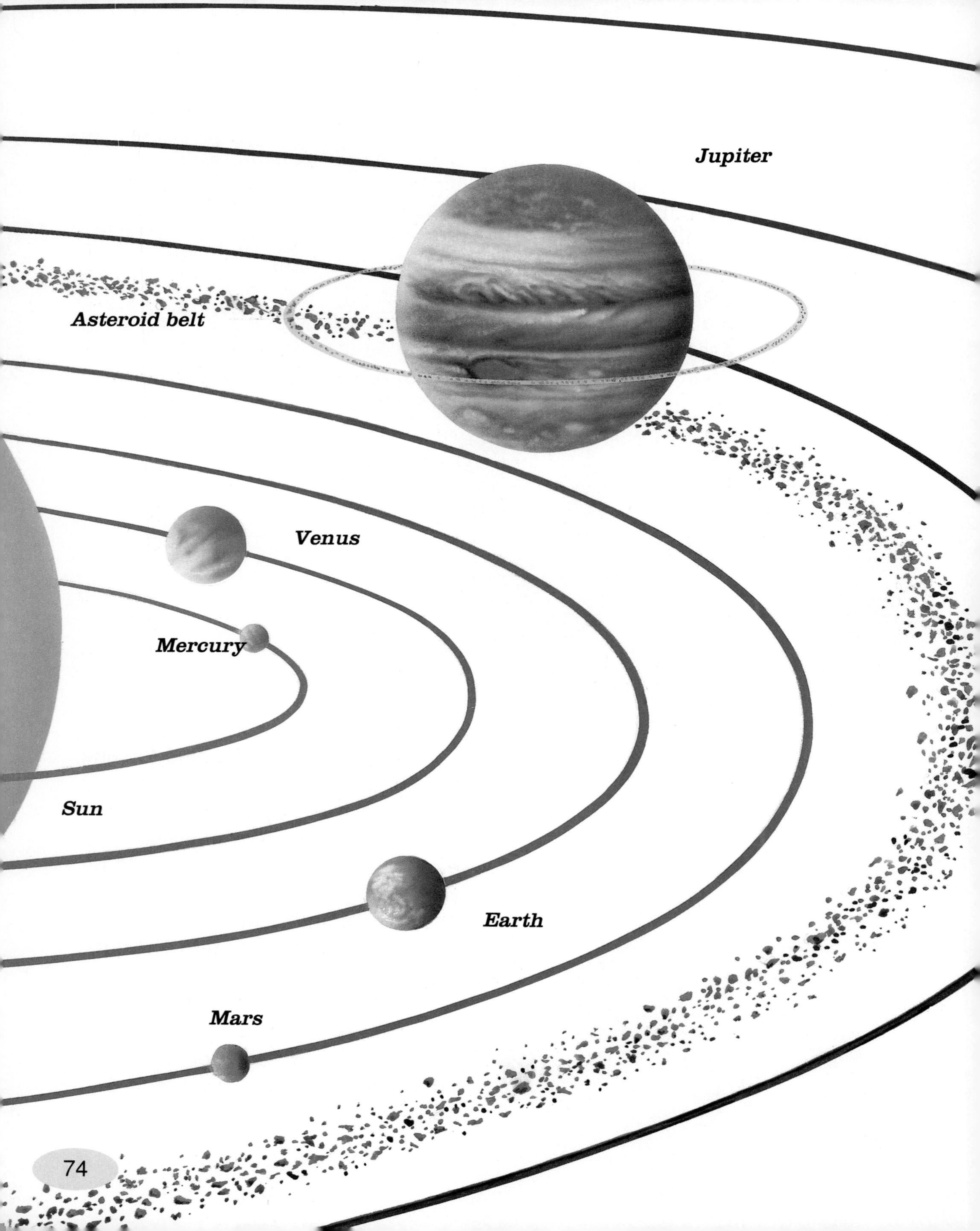
Jupiter
Asteroid belt
Venus
Mercury
Sun
Earth
Mars

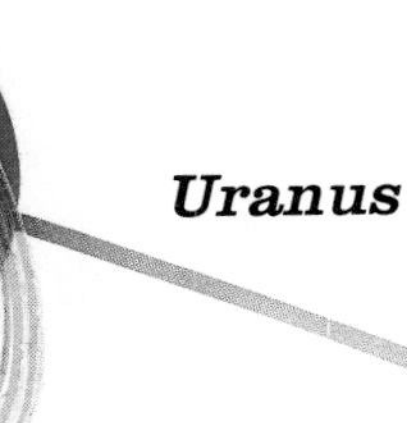

Sum It Up

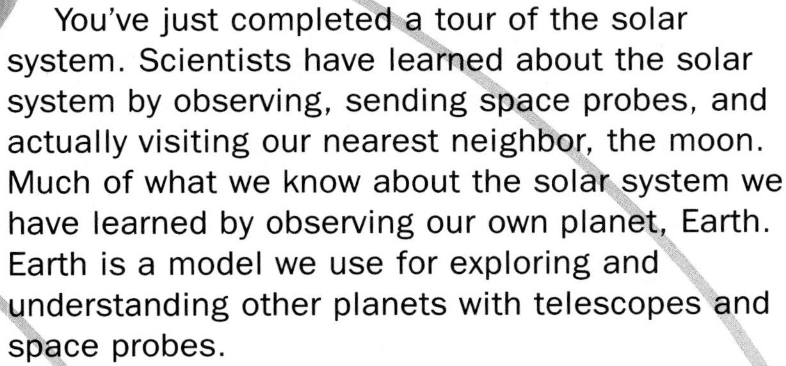

You've just completed a tour of the solar system. Scientists have learned about the solar system by observing, sending space probes, and actually visiting our nearest neighbor, the moon. Much of what we know about the solar system we have learned by observing our own planet, Earth. Earth is a model we use for exploring and understanding other planets with telescopes and space probes.

You learned how the four inner planets in the solar system are made of rock and are different from the five outer ones that are made of gas. You compared the planets in terms of size, distance, and other features. You also examined differences among asteroids, comets, meteoroids, meteors, and meteorites. Scientists are trying to develop models that will explain how the solar system began. They still have many questions to answer. One of the most interesting questions for all of us is whether there are similar planet systems around other stars and whether life exists on them.

Critical Thinking

1. Study the classification of inner planets and outer planets. From the information you have, how would you group the planets? Why?
2. How are "falling stars" different from stars? Could you catch a "falling star?"
3. Why do you think life could or could not exist on other planets in the solar system?

Model of the solar system

Art not to scale

Our Future Is in Space!

In your lifetime we'll learn many new things. Theories and models of space will change. Watch the news, because new probes are on their way to the planets and the sun. Our space shuttles put new satellites into orbit each year. Space Station *Freedom*, a laboratory in the sky, is being designed, and NASA may send astronauts to Mars. Space colonies on the moon and Mars are being planned.

Minds On! On a large sheet of paper, draw a picture showing Earth's place in space. Explain where Earth is in relation to the sun and other planets and to the stars and Milky Way. Compare your drawing with the one you made on page 1 in your ***Activity Log.*** ●

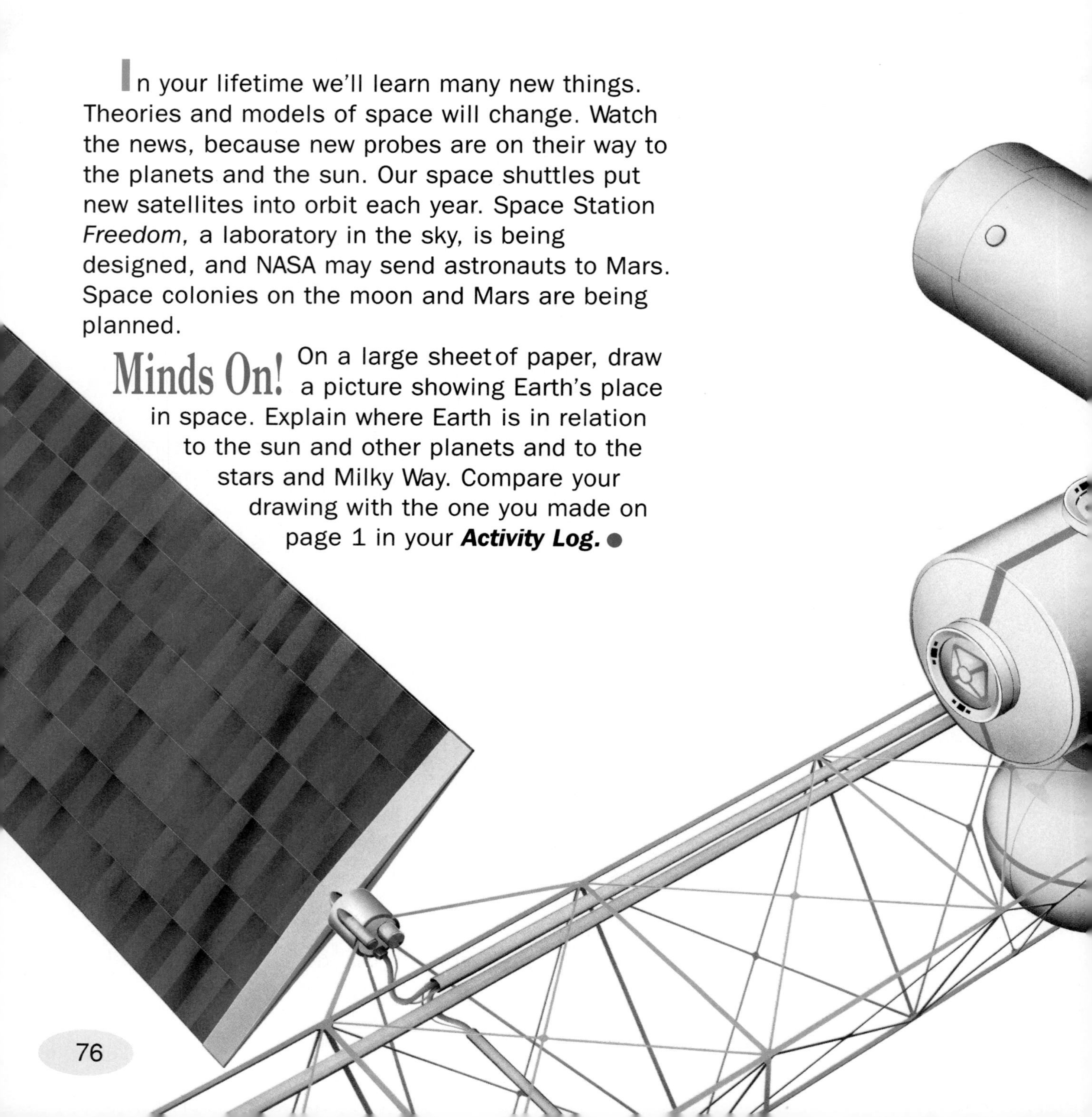

The Macmillan Book of Astronomy

The Macmillan Book of Astonomy has additional information about planets, asteroids, the solar system, sun, stars, and other objects. Use *The Macmillan Book of Astronomy* to find the answer to these questions and many more you may have thought of while you were reading this unit.

- What space probes have been sent to Venus and what did they find?
- What is the sun's chromosphere?
- What is a variable star?
- Why does Mercury have craters?
- Where in the solar system can you find active volcanoes?
- Why are comets called "dirty snowballs?"
- Why is the belt of asteroids and meteoroids where it is?

Vacation Into the Future!

Space exploration is important to our future and to our understanding of both Earth and the universe. In this unit, you've seen some of what students do at Space Camp in Huntsville, Alabama, or Titusville, Florida. Activities at Space Camp include building and launching your own model rocket, tasting space food, trying on space suits, and practicing in the Moon Walk Trainer. You can also see the actual Apollo 16 spacecraft that went to the moon and conduct a simulated space shuttle mission.

Programs for advanced students at Space Camp include space station experiments, jet-flight simulation, and many other activities. They have programs for teachers and other adults, too! You can get your own booklet about U.S. Space Camp by calling 1–800–63 SPACE. U.S. Space Camp is a nonprofit organization cooperating with, but not funded by, NASA.

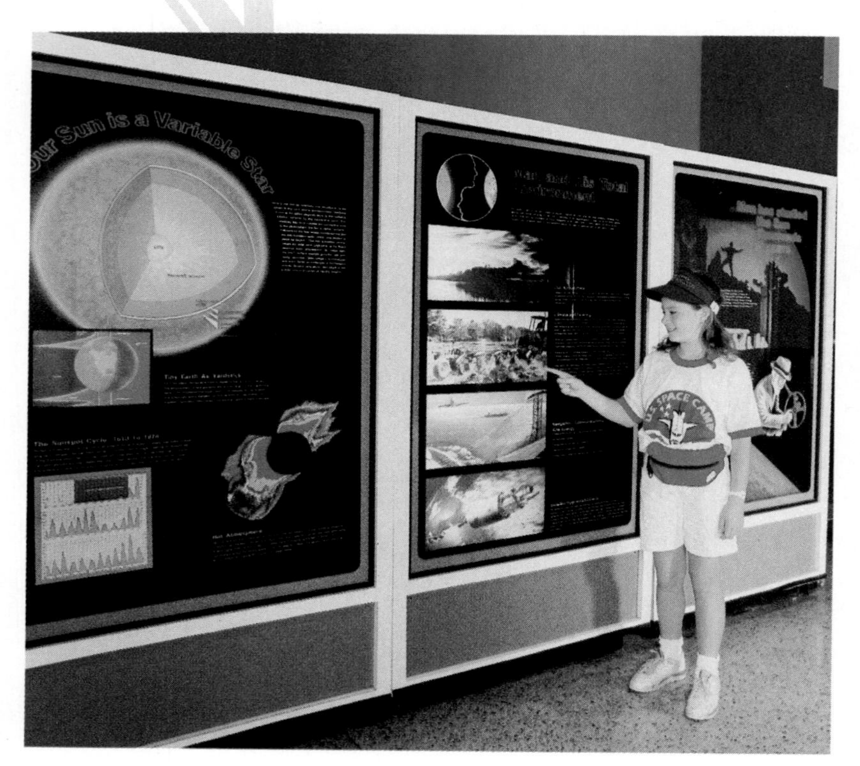

Dimensions in Space is an exhibit in the U.S. Space & Rocket Center museum in Huntsville, Alabama. Here, the general public and U.S. Space Camp and Academy trainees learn about the galaxy, our moon, and exploring the planets in our solar system.

Peaceful Partners to the Planets?

Exploring space is so hard and so expensive that nations are doing it together. The United States and U.S.S.R. shared one mission and will trade data from space medicine research and Mars satellites. NASA and the European Space Agency (13 European nations) together built the Hubble Space Telescope. Nations that work together may get along better.

Does any one of these countries own space? Countries in the United Nations agree that space has no "owner." Like the ocean, it's open for all countries to explore and use.

Cooperation is the key to future space exploration.

Minds On! Extend the time line you created on page 9 in your ***Activity Log*** to include events that you think will happen in the future. Will those events happen soon, or will there be a long period of time before they happen? Why? ●

GLOSSARY

Use the pronunciation key below to help you decode, or read, the pronunciations.

Pronunciation Key

a	**a**t, b**a**d	**d**	**d**ear, so**d**a, ba**d**
ā	**a**pe, p**ai**n, d**ay**, br**ea**k	**f**	**f**ive, de**f**end, lea**f**, o**ff**, cou**gh**, ele**ph**ant
ä	f**a**ther, c**a**r, h**ea**rt	**g**	**g**ame, a**g**o, fo**g**, e**gg**
âr	c**are**, p**air**, b**ear**, th**eir**, wh**ere**	**h**	**h**at, a**h**ead
e	**e**nd, p**e**t, s**ai**d, h**ea**ven, fr**ie**nd	**hw**	**wh**ite, **wh**ether, **wh**ich
ē	**e**qual, m**e**, f**ee**t, t**ea**m, p**ie**ce, k**ey**	**j**	**j**oke, en**j**oy, **g**em, pa**g**e, e**dg**e
i	**i**t, b**i**g, **E**nglish, h**y**mn	**k**	**k**ite, ba**k**ery, see**k**, ta**ck**, **c**at
ī	**i**ce, f**i**ne, l**ie**, m**y**	**l**	**l**id, sai**l**or, fee**l**, ba**ll**, a**ll**ow
îr	**ear**, d**eer**, h**ere**, p**ier**ce	**m**	**m**an, fa**m**ily, drea**m**
o	**o**dd, h**o**t, w**a**tch	**n**	**n**ot, fi**n**al, pa**n**, k**n**ife
ō	**o**ld, **oa**t, t**oe**, l**ow**	**ng**	lo**ng**, si**ng**er, pi**n**k
ô	c**o**ffee, **a**ll, t**au**ght, l**aw**, f**ou**ght	**p**	**p**ail, re**p**air, soa**p**, ha**pp**y
ôr	**or**der, f**or**k, h**or**se, st**or**y, p**our**	**r**	**r**ide, pa**r**ent, wea**r**, mo**r**e, ma**rr**y
oi	**oi**l, t**oy**	**s**	**s**it, a**s**ide, pet**s**, **c**ent, pa**ss**
ou	**ou**t, n**ow**	**sh**	**sh**oe, wa**sh**er, fi**sh** mi**ssi**on, na**ti**on
u	**u**p, m**u**d, l**o**ve, d**ou**ble	**t**	**t**ag, pre**t**end, fa**t**, bu**tt**on, dress**ed**
ū	**u**se, m**u**le, c**ue**, f**eu**d, f**ew**	**th**	**th**in, pan**th**er, bo**th**
ü	r**u**le, tr**ue**, f**oo**d	**<u>th</u>**	**th**is, mo**th**er, smoo**th**
u̇	p**u**t, w**oo**d, sh**ou**ld	**v**	**v**ery, fa**v**or, wa**v**e
ûr	b**ur**n, h**urr**y, t**er**m, b**ir**d, w**or**d, c**our**age	**w**	**w**et, **w**eather, re**w**ard
ə	**a**bout, tak**e**n, penc**i**l, lem**o**n, circ**u**s	**y**	**y**es, on**i**on
b	**b**at, a**b**ove, jo**b**	**z**	**z**oo, la**z**y, ja**zz**, ro**s**e, dog**s**, hous**es**
ch	**ch**in, su**ch**, ma**tch**	**zh**	vi**si**on, trea**s**ure, sei**z**ure

asteroids (as′ tə roids′) fragments of matter similar to rocky planetary matter that orbit between Mars and Jupiter.

astronomy (ə stron′ ə mē) the study of objects in space.

black hole a star in which matter is condensed and its gravity field so strong that light cannot escape.

comet a mass of frozen gases, dust, and small rock particles that orbits the sun.

constellation (kon′ stə lā′ shən) a visual pattern of stars.

galaxy large system of gases, dust, and many stars.

gravity the mutual force of attraction that exists between all objects in the universe.

light-year the distance light travels in one year.

meteor (mē′ tē ər) meteoroids that burn up in Earth's atmosphere.

meteorite (mē′ tē ə rīt′) meteoroids that strike Earth.

meteoriods (mē′ tē ə roidz′) small fragments of matter moving in space that sometimes enter Earth's atmosphere.

nebula (neb′ yə lə) a cloud of dust and gas in which stars may be born.

neutron stars (nū´ tron stärz) a dense mass that results from the collapse of a supernova.

planet an object in space that reflects light from a nearby star around which it revolves.

pulsars (pul´ särz) a rapidly spinning neutron star that sends out radio waves.

satellite (sat´ ə līt´) an object that revolves around a larger body.

sextant an instrument used mainly in navigation for measuring the altitude of the sun or a star to determine the position of the observer.

solar system the system of objects in orbit around our sun.

space probe spacecraft equipped with instruments designed to collect information about outer space.

space shuttle a reusable craft designed to transport astronauts, materials, and satellites to and from space.

sundial a tool that shows the time of the day by the position and length of the shadow the sun casts on a surface marked with numbers.

supernova (sü′ pər nō və) the explosion of a star in which the center of the star collapses under gravity, the outer layers are blown off at high speeds, and the brightness of the star increases greatly before fading.

universe all that exists, including Earth and all of space.

weightlessness being in a state of apparent lack of gravitational pull.

INDEX

CREDITS

Photo Credits:

Cover, The Image Bank/ Joseph Drivas; **3,** (t) © Studiohio, (b) ©Bill Pekala/Peter Arnold, Inc.; **5,** Devaney Stock Photos; **8,** Arni Katz/Unicorn Stock Photo; **9,** Richard Nowitz Photo; **10, 11,** ©Studiohio, Jim Ballard/ALLSTOCK; **12, 13,** ©Bill Pekala/Peter Arnold, Inc.; **14, 15,** ©Studiohio; **18,** (t) ©Comstock Inc., (b) US Space & Rocket Center; **20,** NASA; **21,** NASA/INT'L Stock; **23,** (t) ©Leo Hetzel/Photo Reporters, (b) Frank P. Rossotto; **24,** (t) Photri; (b) NASA; **25,** (t) INT'L Stock, (b) ©Leo DeWys, Inc.; **26,** (t) Richard Nowitz Photo, (b) Photri; **27,** (t) US Space & Rocket Center, (b) David Young Wolff/Photoedit; **28,** (tr) Comstock Inc., (mr) © Brent Turner/BLT Productions/1991, (ml) Image Bank (bl) ©Studiohio, Jim Ballard/ALLSTOCK; **29,** (tr) (tl) ©Leonard Lessin, Peter Arnold, Inc., (ml) Jim Ballard/ALLSTOCK, (mr) (ml) ©KS Studios/1991, (bl) ©Ann Duncan/Tom Stack & Associates; **30,** ©Comstock Inc.; **32, 33,** ©KS Studios/1991; **34,** ©John D. Cunningham/Visuals Unlimited; **43,** ©Russ Kinne/Comstock Inc.; **44,** (i) US Space & Rocket Center, ©David Parker/Photo Researchers; **45,** Carlye Calvin Photos; **48, 49,** ©KS Studios/1991; **50, 53,** ©Visuals Unlimited; **60,** ©Science Source/Photo Researchers; **62, 63,** ©KS Studios/1991; **64,** Photri; **65,** (t) NASA/INT'L Stock, (m) NASA, (b) Photri; **66,** NASA/INT'L Stock; **67,** (t) ©Frank P. Rossotto, (m) Photri; **68,** Culver Pictures; **72,** ©John Cancalosi/Peter Arnold, Inc.; **73,** (t) ©Universal Shooting Star, (b) Photoedit; **78,** US Space & Rocket Center; **79,** Richard Nowitz Photo.

Illustration Credits:

6, 7, Stephanie Pershing; **14, 32, 48, 62,** Bob Giuliani; **16, 19, 22, 36, 74, 75,** Charlie Thomas; **17,** James Shough; **35, 51,** Bill Boyer; **37, 40, 41, 46, 47, 58, 59, 68, 69, 70, 71,** Francisco Maruca; **38, 39, 52, 53, 54,** Bill Singleton; **42, 43, 67, 73,** Anne Rhodes; **55,** Mark Lakin; **56, 57, 76, 77,** Henry Hill